Information Seeking
and Communicating
Behavior
of Scientists and Engineers

Information Seeking and Communicating Behavior of Scientists and Engineers

Cynthia Steinke
Editor

Routledge
Taylor & Francis Group

NEW YORK AND LONDON

First published by:

The Haworth Press, Inc. 10 Alice Street, Binghamton, NY 13904-1580
EUROSPAN/Haworth, 3 Henrietta Street, London WC2E 8LU England

This edition by Routledge:

Routledge
Taylor & Francis Group
270 Madison Avenue
New York, NY 10016

Routledge
Taylor & Francis Group
2 Park Square
Milton Park, Abingdon
Oxon OX14 4RN

Information Seeking and Communicating Behavior of Scientists and Engineers has also been
published as *Science & Technology Libraries*, Volume 11, Number 3 1991.

Library of Congress Cataloging-in-Publication Data

Information seeking and communicating behavior of scientists and engineers / Cynthia Steinke.
editor.
    p. cm.
    "Has also been published as Science & Technology libraries, volume 11, number 3, 1991."
Includes bibliographical references.
    ISBN 1-56024-135-7 (acid-free paper)
    1. Communication in science. 2. Communication of technical information. 3. Scientists. 4.
Engineers. I. Steinke, Cynthia A.
Q223.I49 1991
501.4 – dc20
                                       91-11233
                                         CIP

Information Seeking and Communicating Behavior of Scientists and Engineers

CONTENTS

Meeting the Academic and Research Information Needs of Scientists and Engineers in the University Environment 83
Harry Llull

SPECIAL PAPERS

Physical Structure and Administration of Science and Technology Libraries: An Historical Survey 91
Elizabeth P. Roberts
Elaine Brekke
Kimberly Douglas

Academic Science and Technology Libraries: Facilities and Administration 107
Elaine Brekke
Kimberly Douglas
Elizabeth Roberts

ABOUT THE EDITOR

Cynthia Steinke, MS, is Director of the Institute of Technology Libraries at the University of Minnesota, which provides collections and services for the physical and engineering sciences. Prior to joining the staff at the Institute of Technology Libraries in 1984, she served as Science and Engineering Librarian at the University of Illinois in Chicago, and Science Librarian at the John Crerar Library, also in Chicago.

A member of the Special Libraries Association since 1964, Cynthia Steinke has served as President of the Illinois chapter, Chair of the Science-Technology Division, and member of the National Networking Committee. She has also been active in the American Library Association, the American Society for Information Science, and the International Association of Technological University Libraries.

Introduction

How do scientists and engineers really discover, select and use the countless information and communications resources available to them? This question has been addressed through surveys, studies and analyses for a number of years, with varying degrees of success in identifying characteristic habits and behavior.

What are the implications of advances in computer-based communication technology on information seeking behavior activities, now and in the future? This is the question being asked today.

Despite the abundance of studies, it is generally accepted that we still know far too little about the behavior of our users. As information specialists, we are very knowledgeable about the information systems we manage and use, but are not as familiar with the work, the information needs and the communicating behavior of the research worker.

In telling just a portion of this story, the contributing authors to this volume look at this question from several perspectives: a review of the literature of the past three decades to the "cold fusion furor" that had scientists and the press bubbling over like test tubes in the early months of 1989.

Leading off, Pinelli (NASA Langley Research Center) discusses the differences between scientists and engineers—the relationships between science and technology. His paper then focuses on the information seeking behavior of engineers, highlighting major studies that have been conducted, followed by a discussion of the mandate we have before us to better understand the habits and practices of engineers, if we are to develop information systems that will truly meet their needs.

Robert Allen (Purdue University, Physics Library) looks at the communication patterns of scientists in general, and physicists in particular, which are undergoing change due to the introduction of

1

new electronic modes of communication. Bichteler (University of Texas, Austin) discusses the gray literature that geologists depend *upon for unique information on local and regional geology,* despite the fact that this material has typically been excluded from conventional channels of library acquisitions and processing. Librarians find these "sneaky, fly-by-night, change-coat publications hard to identify, hard to acquire, hard to catalog and retrieve and hard to preserve . . . a librarian's ultimate challenge." Yet, this material may well be more critical to users than material which is given full processing.

Using the story of cold fusion research, Welborn (University of California, Santa Cruz) re-examines the role of the peer review function as a traditional element in the information seeking/communication cycle of research scientists.

Poland (Purdue University, Engineering Library) in her literature review offers a view of the magnitude and findings of studies that have been conducted during the past thirty years in an effort to capture the essense of communication behavior among scientists. Effects of new communication technologies are also considered by the author.

Two shorter pieces conclude this discussion of information seeking habits of researchers. Sieving (University of Michigan Medical Center) offers a delightful musing on the types of information seekers we have all known (and loved) during the course of our careers as informational professionals. Llull (University of New Mexico) describes how the organizational structures and information services of academic research libraries are being challenged to respond to the new information needs of the faculty and researchers, especially as provoked by the tremendous growth in research and development funding awarded by the federal government to universities and research centers.

A common theme runs through all the selections offered in this volume. Despite an abundance of studies, we still don't understand the information seeking habits of our user communities well enough and thus, we probably are not meeting them. In addition, different fields have developed varying systems of communication which must be identified and recognized.

What is known about the information practices of researchers has not been applied to existing information services. Pinelli and others point out that information professionals tend to place more emphasis on the technology instead of the quality of the information itself and its ability to meet the needs of the user. At the same time, communication styles are changing, largely as a result of new technologies.

The message? We must make an ever greater effort to understand the work of our users, the differing information channels and sources they employ, and thereby tailor our systems and services to support their information seeking behavior. We need to better understand how advances in computer and communications technologies will influence and change the utilization of resources; how the new technologies can more effectively be used to facilitate the communication process.

* * *

The Sci-Tech Collection Development paper deals with the difficult question of radioactive waste management. In their review, Cromer and Thomas provide an overview of the subject as well as the basic literature sources for building collections and working with users who seek information in this area. Largely focused on sci-tech literature, some discussion of the social and political issues is provided as well.

A recent survey of seventy-five Association of Research Libraries provided data to analyze the structures of sci/tech libraries. Two companion papers (Brekke, Roberts and Douglas) compare the physical and administrative structures of centralized, multiple branch/departmental and integrated (collections merged into a main library facility) sci-tech libraries. The evolution of administrative structures during the past 40 years is discussed, along with some forecast of future trends.

Cynthia A. Steinke
Editor

NOTES

1. *Communications in Support of Science and Engineers; a report to the National Science Foundation from the Council on Library Resources*. Washington, D.C., Council on Library Resources, August, 1990.
2. Bishop, Ann P.; Peterson, Marshall B. Developing Information Systems for Technology Transfer: An example in tribology. *Science and Technology Libraries* 11 (2), Winter, 1990.

The Information-Seeking Habits and Practices of Engineers

Thomas E. Pinelli

INTRODUCTION

There are many different information user communities. The differences between them may be great. Even within similar or related user communities there may be considerable differences among users. Thus, to meet the information needs of the user communities, information professionals must first understand the nature of the user community and become familiar with the information-seeking habits and practices of the user. Generally, this has not been the case in science and technology. Information professionals have assumed certain similarities between science and technology and scientists and engineers.

The two communities and user groups are not the same and the argument that a scientist is a more generic term merely evades the fundamental issue. The practice of lumping the two groups [engineers and scientists] together is self-defeating in information [production, transfer, and] use studies because confusion over the characteristics of the sample has led to what appears to be conflicting results and to a greater difficulty in developing normative measures for improving information systems in either science or technology.

Further, the terms engineer and scientist are not synonymous. The difference in work environment and personal/professional goals between the engineer and scientist proves to be an important factor in determining their information-seeking habits and practices. This review paper explores the science/technology and scientist/engineer dichotomy and focuses on the information-seeking habits and practices of the engineer.

Thomas E. Pinelli is affiliated with NASA Langley Research Center, Hampton, VA 23665.

BACKGROUND

In their treatise, *The Positive Sum Strategy: Harnessing Technology for Economic Growth*, Landau and Rosenberg[22] describe technological innovation as *the* critical factor in the long-term economic growth of modern industrial societies that functions successfully *only* within a larger social environment that provides an effective combination of incentives and complementary inputs into the innovation process. Technological innovation is a process in which the communication of STI is critical to the success of the enterprise.[12,35]

"*Technology*, unlike science, is an extroverted activity; it involves a search for workable solutions to problems. When it finds solutions that are workable and effective, it does not pursue the *why?* very hard. Moreover, the output of technology is a product, process, or service. *Science*, by contrast, is an introverted activity. It studies problems that are usually generated internally by logical discrepancies or internal inconsistencies or by anomalous observations that cannot be accounted for within the present intellectual framework."[22] Technology is a process dominated by engineers, as opposed to scientists, which "leads to different philosophies and habits not only about contributing to the technical literature but also to using the technical literature and other sources of information."[17] Consequently, an understanding of the relationship between science and technology and the information-seeking habits and practices of engineers is essential to the development and provision of information services for engineers.

THE NATURE OF SCIENCE AND TECHNOLOGY

The relationship between science and technology is often expressed as a continuous process or normal progression from basic research (science) through applied research (technology) to development (utilization). This relationship is based on the widely held assumption that technology grows out of or is dependent upon science for its development. However, the belief that technological change is somehow based on scientific advance has been challenged in recent years. Substantial evidence exists that refutes the relationship between science and technology.

Schmookler[12] has attempted to show that the variation in inventive activity between different American industries is explicable in terms of the variation in demand, concluding that economic growth determines the rate of inventive activity rather than the reverse. Price,[25] in his investigation of citation patterns in both scientific and technical journals, found that scientific literature is cumulative and builds upon itself, whereas technical literature is not and does not build upon itself. Citations to previous work are fewer in technical journals and are often the author's own work.

Price[25] concluded that science and technology progress independently of one another. Technology builds upon its own prior developments and advances in a manner independent of any link with the current scientific frontier and often without any necessity for an understanding of the basic science underlying it.

In summarizing the differences between science and technology, Price[25] makes the following 12 points. *First*, science has a cumulating, close-knit structure; that is, new knowledge seems to flow from highly related and rather recent pieces of old knowledge, as displayed in the literature. *Second*, this property is what distinguishes science from technology and from humanistic scholarship. *Third*, this property accounts for many known social phenomena in science and also for its surefootedness and high rate of exponential growth. *Fourth*, technology shares with science the same high growth rate, but shows quite complementary social phenomena, particularly in its attitude to the literature. *Fifth*, technology therefore may have a similar, cumulating, close-knit structure to that of science, but it is of the state of the art rather than of the literature. *Sixth*, science and technology each therefore have their own separate cumulating structures. *Seventh*, a direct flow from the research front of science to that of technology, or vice versa, occurs only in special and traumatic cases since the structures are separate.

Eighth, it is probable that research-front technology is strongly related only to that part of scientific knowledge that has been packed down as part of ambient learning and education, not to research-front science. *Ninth*, research-front science is similarly related only to the ambient technological knowledge of the previous generation of students, not to the research front of the technological state of the art and its innovation. *Tenth*, this reciprocal relation between science and technology, involving the research front of one and the accrued archive of the other, is nevertheless sufficient to keep the two in phase in their separate growths within each otherwise independent cumulation. *Eleventh*, it is therefore naive to regard technology as applied science or clinical practice as applied medical science. *Twelfth*, because of this, one should be aware of any claims that a particular scientific research is needed for particular technological breakthroughs, and vice versa. Both cumulations can only be supported for their own separate ends.

Allen,[2] who studied the transfer of technology and the dissemination of technological information in R&D organizations, finds little evidence to support the relationship between science and technology as a continuous relationship. Allen concludes that the relationship between science and technology is best described as a series of interactions that are based on need rather than on a normal progression.

Allen[2] states that the independent nature of science and technology (S&T) and the different functions performed by engineers and scientists directly influence the flow of information in science and technology. Science and technology are ardent consumers of information. Both engineers and scientists require large quantities of information to perform their work. At this level, there is a strong similarity between the information input needs of engineers and scientists. However, the difference between engineers and scientists in terms of information processing becomes apparent upon examination of their outputs.[2]

According to Allen[2] information processing in S&T is depicted in the form of an input-output model. Scientists use information to produce information. From a system standpoint, the input and output, which are both verbal, are compatible. The output from one stage is in a form required for the next stage. Engineers use information to produce some physical change in the world. Engineers consume information, transform it, and produce a product that is information bearing; however, the information is no longer in verbal form. Whereas scientists consume and produce information in the form of human language, engineers transform information from a verbal format to a physically encoded form. Verbal information is produced only as a by-product to document the hardware and other physical products produced.

According to Allen,[2] there is an inherent compatibility between the inputs and outputs of the information-processing system of science. He further states that since both are in a verbal format, the output of one stage is in the format required for the next stage. The problem of supplying information to the scientist becomes a matter of collecting and organizing these outputs and making them accessible. Since science operates for the most part on the premise of free and open access to information, the problem of collecting outputs is made easier.

In technology, however, there is an inherent incompatibility between inputs and outputs. Since outputs are usually in a form different from inputs, they usually cannot serve as inputs for the next stage. Further, the outputs are usually in two parts, one physically encoded and the other verbally encoded. The verbally encoded part usually cannot serve as input for the next stage because it is a by-product of the process and is itself incomplete.[2] Those unacquainted with the development of the hardware or physical product therefore require some human intervention to supplement and interpret the information contained in the documentation.[1] Since technology operates to a large extent on the premise of restricted access to information, the problem of collecting the documentation and obtaining the necessary human intervention becomes difficult.[12]

DISTINGUISHING ENGINEERS FROM SCIENTISTS

In their study of the values and career orientation of engineering and science undergraduate students, Krulee and Nadler[21] found that engineering and science students have certain aspirations in common: to better themselves and to achieve a higher socioeconomic status than that of their parents. They reported that science students place a higher value on independence and on learning for its own sake while engineering students are more concerned with success and professional preparation. Many engineering students expect their families to be more important than their careers as a source of satisfaction, but the reverse pattern is more typical for science students.

Krulee and Nadler[21] also determined that engineering students are less concerned than science students with what one does in a given position and more concerned with the certainty of the rewards to be obtained. They reported that, overall, engineering students place less emphasis on independence, career satisfaction, and the inherent interest their specialty holds for them and place more value on success, family life, and avoiding a low-level job. Engineering students appear to be prepared to sacrifice some of their independence and opportunities for innovation in order to realize their primary objectives. Engineering students are more willing to accept positions that will involve them in complex organizational responsibilities and they assume that success in such positions will depend upon practical knowledge, administrative ability, and human relation skills.[21]

In his study of engineers in industry, Ritti[27] found marked contrast between the work goals of engineers and scientists. Ritti draws the following three conclusions from his study: (1) the goals of engineers in industry are very much in line with meeting schedules, developing products that will be successful in the marketplace, and helping the company expand its activities; (2) while both engineers and scientists desire career development or advancement, for the engineer advancement is tied to activities within the organization, while advancement for the scientist is dependent upon the reputation established outside of the organization; and (3) while publication of results and professional autonomy are clearly valued goals of the Ph.D. scientist, they are clearly the least valued goals of the baccalaureate engineer.

Allen[1] states that the type of person who is attracted to a career in engineering is fundamentally different from the type of person who pursues a career as a scientist. He writes that "perhaps the single most important difference between the two is the level of education. Engineers are generally educated to the baccalaureate level; some have a master's degree while some have no college degree. The research scientist is usually assumed to have a doctorate. The long, complex process of academic socialization involved in obtaining the Ph.D. is bound to result in persons who differ considerably in their lifeviews." According to Allen,[1] these differences in values and attitudes toward work will almost certainly be reflected in the behavior of the individual, especially in their use and production of information.

According to Blade,[6] engineers and scientists differ in training, values, and methods of thought. Further, Blade states that the following differences exist in their individual creative processes and in their creative products: (1) scientists are concerned with discovering and explaining nature; engineers use and exploit nature; (2) scientists are searching for theories and principles; engineers seek to develop and make things; (3) scientists are seeking a result for its own ends; engineers are engaged in solving a problem for the practical operating results; and (4) scientists create new unities of thought; engineers invent things and solve problems. Blade states that "this is a different order of creativity."

Finally, communication in engineering and science are fundamentally different. Communication patterns differ because of the fundamental differences between engineering and science and because of the social systems associated with the two disciplines. With one exception, the following characteristics of the social systems as they apply to the engineer and scientist are based on Holmfeld's[16] investigation of the communication behavior of engineers and scientists.

Engineer

- Contribution is [technical] knowledge used to produce end-items or products.
- New and original knowledge is not a requirement.
- Reward is monetary or materialistic and serves as an inducement to continue to make further contributions to technical knowledge.
- Seeking rewards that are not part of the social system of technology is quite proper and also encouraged.
- The value of technical knowledge lies in its value as a commodity of indirect exchange.
- Exchange networks found in the social system of technology are based on end-item products, not knowledge.
- Strong norms against free exchange or open access to knowledge with others outside of the organization exist in the social system of technology.
- Restriction, security classification, and proprietary claims to knowledge characterize the social system of technology.

Scientist

- Contribution is new and original knowledge.
- Reward is social approval in the form of professional [collegial] recognition.
- Recognition is established through publication and claim of discovery.
- A well-developed communication system based on unrestricted access is imperative to recognition and claim of discovery.
- Since recognition and priority of discovery are critical, strong norms against any restriction to free and open communication exist in the social system of science.
- Seeking rewards that are not part of the social system of science in return for scientific contribution is not considered proper within the social system of science.
- Exchange networks commonly referred to as "invisible colleges" exist in the social system of science; in these networks the commodities are knowledge and recognition.[26,10]

INFLUENCE ON INFORMATION-SEEKING HABITS AND PRACTICES OF ENGINEERS

The nature of science and technology and differences between engineers and scientists influence their information-seeking habits, practices, needs, and preferences and have significant implications for planning information services for these two groups.[36] Taylor,[37] who quotes Brinberg,[7] offers the following characteristics for engineers and scientists: "Unlike scientists, the goal of the engineer is to produce or design a product, process, or system; not to publish and make original contributions to the literature. Engineers, unlike scientists, work within time constraints; they are not interested in theory, source data, and guides to the literature nearly so much as they are in reliable answers to specific questions. Engineers prefer informal sources of information, especially conversations with individuals *within* their organization. Finally, engineers tend to minimize loss rather than maximize gain when seeking information."

Anthony et al.,[4] suggest that engineers may have psychological traits that predispose them to solve problems alone or with the help of colleagues rather than finding answers in the literature. They further state that "engineers like to solve their own problems. They draw on past experiences, use the trial and error method, and ask colleagues known to be efficient and reliable instead of searching or having someone search the literature for them. They are highly independent and self-reliant without being positively antisocial."

According to Allen,[2] "Engineers read less than scientists, they use literature and libraries less, and seldom use information services which are directly oriented to them. They are more likely to use specific forms of literature such as handbooks, standards, specifications, and technical reports." What an engineer usually wants, according to Cairns and Compton,[8] is "a specific answer, in terms and format that are intelligible to him — not a collection of documents that he must sift, evaluate, and translate before he can apply them." Young and Harriott[38] report that "the engineer's search for information seems to be based more on a need for specific problem solving than around a search for general opportunity. When engineers use the library, it is more in a personal-search mode, generally not involving the professional (but "nontechnical") librarian." Young and Harriot conclude by saying that "when engineers need technical information, they usually use the most accessible sources rather than searching for the highest quality sources. These accessible sources are respected colleagues, vendors, a familiar but possibly out dated text, and internal company [technical] reports. He [the engineer] prefers informal information networks to the more formal search of publicly available and cataloged information."

Evidence exists to support the hypothesis that differences between science and technology and scientists and engineers directly influence information-seeking habits, practices, needs, and preferences. The results of a study conducted by the System Development Corporation[16] determined that "an individual differs systematically from others in his use of STI" for a variety of reasons. Chief among these are five institutional variables — type of researcher, engineer or scientist; type of discipline, basic or applied; stage of project, task, or problem completeness; the kind of organization, fundamentally thought of as academia, government, and industry; and the years of professional work experience."

Studies, such as the work performed by O'Gara,[23] indicate that information-seeking habits, practices, needs, and preferences are influenced by certain sociometric variables. O'Gara found a positive correlation between physical proximity to an information source and its use. King et al.,[19] report a positive correlation between the number of visits to a library and proximity of the user.

Gerstberger and Allen,[13] in their study of engineers and their choice of an information channel, note the following:

> Engineers, in selecting among information channels, act in a manner which is intended not to maximize gain, but rather to minimize loss. The loss to be minimized is the cost in terms of effort, either physical or psychological, which must be expended in order to gain access to an information channel.

Their behavior appears to follow a "law of least effort."[19] According to this law, individuals, when choosing among several paths to a goal, will base their decision upon the single criterion of "least average rate of probable work." According to Gerstberger and Allen, engineers appear to be governed or influenced by a principle closely related to this law. They attempt to minimize effort in terms of work required to gain access to an information channel. Gerstberger and Allen reached the following conclusions:

1. Accessibility is the single most important determinant of the overall extent to which an information channel or source is used by an engineer.
2. Both accessibility and perceived technical quality influence the choice of the first source.
3. The perception of accessibility is influenced by experience. The more experience engineers have with an information channel or source, the more accessible they perceive it to be.

Gerstberger and Allen[13] conclude their discussion by stating that "any assumption that engineers act in accord with a simple instrumental learning model in which they turn most frequently to those information channels which reward them most often should now clearly be laid to rest." Rosenberg's[28] findings also support the conclusions by Gerstberger and Allen that accessibility almost exclusively determines the frequency of use of information channels. Rosenberg concludes that researchers minimize the cost of obtaining information while sacrificing the quality of the information received. In his study of the *Factors Related to the Use of Technical Information in Engineering Problem Solving,* Kaufman[18] reports that engineers rated *technical quality* or *reliability* followed by *relevance* as the criteria used in choosing the most useful information source. However, *accessibility* appears to be used most often for selecting an information source *even if that source* proved to be the least useful.

THE INFORMATION-SEEKING HABITS AND PRACTICES OF ENGINEERS

Studies specifically concerned with the information-seeking habits and practices of engineers were reviewed to further develop the conceptual framework for this paper. Research studies deemed significant are listed in the "Overview of Engineering STI Studies" and are discussed in some detail. Although not comprehensive, data from these studies are used to further develop the concept of "different" information-seeking habits and practices for engineers and scientists. (See Figure 1.)

Herner

Herner's[15] is one of the first "user" studies that is specifically concerned with "differences" in information-seeking habits and practices. He reports significant differences in terms of researchers performing "basic and applied" research, researchers performing "academic and industry" type duties, and their information-seeking habits and practices. Herner says that researchers performing "basic or academic" duties make greater use of formal information channels or sources, depend mainly on the library for their published material, and maintain a significant number of contacts outside of the organization.

Researchers performing "applied or industry" duties make greater use of informal channels or sources, depend on their personal collections of information and colleagues for information,

FIGURE 1. Overview of Engineering STI Studies

Year	Principal Investigator	Research Method	Population	Sample Frame	Sample Design	Sample Size	Percentage Response Rate (number responding)	Description
1954	Herner	Structured interview	All scientific and technical personnel at Johns Hopkins	Unknown	Unknown	600	100	Survey to determine the information-gathering methods of scientific and technical personnel at Johns Hopkins
1970	Rosenbloom and Wolek	Self-administered questionnaire	Members of 5 industrial R&D organizations	2430	Census	2430	71 (1735)	Survey to determine how engineers and scientists in industrial research and development organizations acquire STI
			Members of 4 IEEE interest groups	Unknown	Probability	Unknown	Unknown (1034)	
1977	Allen	Record analysis Self-administered questionnaire	Unknown	Unknown	Unknown	Unknown	Unknown (1153)	Survey to determine technology transfer and the dissemination of technological information in research and development organizations
1980	Kremer	Self-administered questionnaire	All design engineers at one engineering design firm	73	Census	73	82 (60)	Survey to identify and evaluate the information channels used by engineers in a design company
1981	Shuchman	Structured interview Self-administered questionnaire	Engineers in 89 R&D and non-R&D organizations	14797	Probability	3371	39 (1315)	Survey to determine information used and production in engineering
1983	Kaufman	Self-administered questionnaire	Engineers in six technology-based organizations	147	Census	147	100 (147)	Survey to determine the use of technical information in technical problem solving

make significantly less use of the library than do their counterparts, and maintain fewer contacts outside of the organization. Applied or industry researchers make substantial use of handbooks, standards, and technical reports. They also read less and do less of their reading in the library than do their counterparts.[15]

Rosenbloom and Wolek

Rosenbloom and Wolek[29] conducted one of the first "large-scale" industry studies that was specifically concerned with the flow of STI within R&D organizations. They report three significant and fundamental differences between engineers and scientists:

(1) engineers tend to make substantially greater use of information sources *within* the organization than do scientists; (2) scientists make considerably greater use of the professional (formal) literature than do engineers; and (3) scientists are more likely than engineers to acquire information as a consequence of activities directed toward general competence rather than a specific task.

In terms of interpersonal communication, the engineers in the Rosenbloom and Wolek (1970) study recorded a higher incidence of interpersonal communication with people in other parts of their own corporation, whereas scientists recorded a greater incidence of interpersonal communication with individuals employed outside their own corporation. When using the literature, the engineers tend to consult in-house technical reports or trade publications, while the scientists tend to make greater use of the professional [formal] literature.

Rosenbloom and Wolek[20] report certain similarities between engineers and scientists as follows. The propensity to use alternative types of technical information sources is related to the purposes that will give meaning to the use of that information. Work that has a professional focus draws heavily on sources of information external to the user's organization. Work that has an operational focus seldom draws on external sources, relying heavily on information that is available within the employing organization. Those engineers and scientists engaged in professional work commonly emphasize the simplicity, precision, and analytical or empirical rigor of the information source. Conversely, those engineers and scientists engaged in operational work typically emphasize the value of communication with others who understand and are experienced in the same real context of work.

Allen

Allen's[2] study of technology transfer and the dissemination of technological information within the R&D organization is the result of a 10-year investigation. Allen describes the study, which began as a "user study," as a systems-level approach to the problem of communication in technology. Allen's work is considered by many information professionals to be the seminal work on the flow of technical information within R&D organizations.

Allen[2] was among the first to produce evidence supporting different information-seeking habits and practices for engineers and scientists. These differences, Allen notes, lead to different philosophies and habits regarding the use of the technical literature and other sources of information by engineers. The most significant of Allen's findings is the relative lack of importance of the technical literature in terms of generating new ideas and in problem definition, the importance of personal contacts and discussions between engineers, the existence of technological "gatekeepers," and the importance of the technical report. Allen states that "the unpublished report is the single most important informal literature source; it is the principal written vehicle for transferring information in technology."

Kremer

Kremer's[20] study was undertaken to gain insight on how technical information flows through formal and informal channels among engineers in a design company. The engineers in her study are not involved in R&D. Kremer's findings are summarized as follows.

In terms of information habits, personal files are the most frequently consulted source for needed information. The reason given most frequently to search for information is problem solving; colleagues within the company are contacted first for needed information, followed by colleagues outside of the company. In terms of the technical literature, handbooks are considered most important, followed by standards and specifications. Libraries are not considered to be important sources of information and are used infrequently by company engineers.

Regardless of age and work experience, design engineers demonstrate a decided preference for internal sources of information. The perceived accessibility, ease of use, technical quality, and amount of experience a design engineer has had with an information source strongly influence the selection of an information source. Technological gatekeepers were found to exist among design engineers;

they are high technical performers and a high percentage are first line supervisors.

Shuchman

Shuchman's[33] study is a broad-based investigation of information transfer in engineering. The respondents represented 14 industries and the following major disciplines: civil, electrical, mechanical, industrial, chemical and environmental, and aeronautical. Seven percent, or 93 respondents, were aeronautical engineers. The engineers in Shuchman's study, regardless of discipline, display a strong preference for informal sources of information. Further, these engineers rarely find all the information they need for solving technical problems in one source; the major difficulty engineers encounter in finding the information they need to do their job is identifying a specific piece of missing data and then learning who has it.

In terms of information sources and solving technical problems, Shuchman (1981) reports that engineers first consult their personal store of technical information, followed in order by informal discussions with colleagues, discussions with supervisors, use of internal technical reports, and contact with a "key" person in the organization who usually knows where the needed information may be located. Shuchman stated that technical libraries and librarians are used by a small proportion of the engineering profession.

In general, Shuchman[33] states engineers do not regard information technology as an important adjunct to the process of producing, transferring, and using information. While technological gatekeepers appear to exist across the broad range of engineering disciplines, their function and significance is not uniform; considering the totality of engineering, gatekeepers account for only a small part of the information transfer process.

Kaufman

Kaufman's[18] study is concerned with the factors relating to the use of technical information by engineers in problem solving. The study reported that, in terms of information sources, engineers consult their personal collections first, followed by colleagues and then by formal literature sources. In terms of formal literature sources used for technical problem solving, engineers use technical reports, followed in order by books, monographs, and technical handbooks.

Most sources of information, according to Kaufman,[18] are found primarily through an intentional search of written information, followed by personal knowledge and then by asking someone. The study purports that the criteria used in selecting all information sources, in descending order of frequency, are accessibility, familiarity or experience, technical quality, relevance, comprehensiveness, ease of use, and expense. The various information sources are used by engineers for specific purposes. Librarians and information specialists are primarily utilized to find leads to information sources. On-line computer searches are used primarily to define the problem. The technical literature is used primarily to learn techniques applicable to dealing with the problem. Personal experience is used primarily to find solutions to the problem. Kaufman[18] reports that the criteria used in selecting the most useful information sources, in descending order of frequency, are technical quality or reliability, relevance, accessibility, familiarity or experience, comprehensiveness, ease of use, and expense. In terms of the effectiveness, efficiency, and usefulness of the various information sources, personal experience is rated as the most effective in accomplishing the purpose for which it is used; librarians and information specialists receive the lowest rating for efficiency and effectiveness. Most engineers use several different types of information sources in problem solving; however, engineers do depend on their personal experience more often than on any single specific information source.

DISCUSSION

The ability of engineers to identify, acquire, and utilize scientific and technical information (STI) is of paramount importance to the efficiency of technological innovation and the R&D process. Testimony to the central role of STI in the R&D process is found in numerous studies. These studies show, among other things, that engineers and scientists, and aerospace engineers and scientists in particular, devote more time, on the average, to the communication of technical information than to any other scientific or technical activity.[12,24]

A number of studies have found strong relationships between the communication of STI and technical performance at both the individual[3,14,30] and group levels.[9,31,34] As Fischer[12] concludes, "The role of scientific and technical communication is thus central to the success of the innovation process, in general, and the management of R&D activities, in particular." But as Solomon and Tornatzky[35] point out, "While STI, its transfer and utilization, is crucial to technological innovation [and competitiveness], linkages between [the] various sectors of the technology infrastructure are weak and/or poorly defined."

Economists, such as David,[11] point out that "technological innovation is the primary, if not the only means of improving industrial productivity. It is the force propelling the American economy forward and a process [that is] inextricably linked to knowledge transfer and diffusion." In spite of its importance to the U.S. economy and the balance of trade, very little is known about technological innovation and the diffusion of knowledge both in terms of the channels used to communicate the ideas and the information-gathering habits and practices of the members of the social system involved in technological innovation.

Therefore, it is likely that an understanding of the process by which STI is communicated through certain channels over time among the members of the technological social system would contribute to stimulating technological innovation, maximizing the R&D process, increasing R&D productivity, and improving and maintaining the professional competence of U.S. engineers.

However, despite the vast amount of STI available to potential users, several major barriers to effective knowledge diffusion exist.[5] *First*, the very low level of support for knowledge transfer in comparison to knowledge production suggests that dissemination efforts are not viewed as an important component of the R&D process. *Second*, there are mounting reports from users about difficulties in getting appropriate information in forms useful for problem solving and decision making. *Third*, rapid advances in many areas of S&T knowledge can be fully exploited only if they are quickly translated into further research and application. Although the United States dominates basic R&D, foreign competitors may be better able to apply the results. *Fourth*, current mechanisms are often inadequate to help the user assess the quality of available information. *Fifth*, the characteristics of actual usage behavior are not sufficiently taken into account in making available useful and easily retrieved information. These deficiencies must be remedied if the results of R&D are to be successfully applied to innovation, problem solving, and productivity.

CONCLUDING REMARKS

Only by maximizing the R&D process can the United States maintain and possibly recapture its international competitive edge. Key to this goal is the provision of information services and products which meet the information needs of engineers. Evidence exists which indicates that traditional information services and products may, in fact, not be meeting the information needs of engineers. The primary reason(s) for this deficiency is three fold. First, the specific information needs of engineers are neither not well known nor well understood. Second, what is known about the information-seeking habits and practices of engineers has not been applied to existing engineering information services. Third, information professionals continue to over emphasize technology instead of concentrating on the quality of the information itself and the ability of the information to meet the needs of the user.

REFERENCES

1. Allen, Thomas J. "Distinguishing Engineers From Scientists." In *Managing Professionals in Innovative Organizations*, Ralph Katz, ed. (Cambridge, MA: Ballinger Publishing, 1988,) 3-18.

2. Allen, Thomas J. *Managing the Flow of Technology: Technology Transfer and the Dissemination of Technological Information Within the R&D Organization*. (Cambridge, MA: MIT Press, 1977.)

3. Allen, Thomas J. "Roles in Technical Communication Networks." In *Communication Among Scientists and Engineers*, Carnot E. Nelson and Donald K. Pollack, eds. (Lexington, MA: D.C. Heath, 1970,) 191-208.

4. Anthony, L. J.; H. East; and M. J. Slater. "The Growth of the Literature of Physics." *Reports on Progress in Physics* 32 (1969): 709-767.

5. Bikson, Tora K.; Barbara E. Quint; and Leland L. Johnson. *Scientific and Technical Information Transfer: Issues and Options*. (Washington, DC: National

Science Foundation, March 1984.) (Available from NTIS, Springfield, VA PB-85-150357; also available as Rand Note 2131.)

6. Blade, Mary-Frances. "Creativity in Engineering." In *Essays on Creativity in the Sciences*. Myron A. Coler, ed. (NY: New York University Press, 1963,) 110-122.

7. Brinberg, Herbert R. "The Contribution of Information to Economic Growth and Development." Paper presented at the 40th Congress of the International Federation for Documentation, Copenhagen, Denmark, August 1980.

8. Cairns, R.W. and Bertita E. Compton. "The SATCOM Report and the Engineer's Information Problem." *Engineering Education* 60:5 (January 1970): 375-376.

9. Carter, C.F. and B.R. Williams. *Industry and Technical Progress: Factors Governing the Speed of Application of Science*. (London: Oxford University Press, 1957.)

10. Crane, Diana. *Invisible Colleges: Diffusion of Knowledge in Scientific Communities*. (Chicago: University of Chicago Press, 1972.)

11. David, Paul A. "Technology Diffusion, Public Policy, and Industrial Competitiveness." In *The Positive Sum Strategy: Harnessing Technology for Economic Growth*, Ralph Landau and Nathan Rosenberg, eds. (Washington, DC: National Academy Press, 1986,) 373-391.

12. Fischer, William A. "Scientific and Technical Information and the Performance of R&D Groups." In *Management of Research and Innovation*, Burton V. Dean and Joel L. Goldhar, eds. (NY: North-Holland Publishing Company, 1980,) 67-89.

13. Gerstberger, Peter G. and Thomas J. Allen. "Criteria Used By Research and Development Engineers in the Selection of an Information Source." *Journal of Applied Psychology* 52:4 (August 1968): 272-279.

14. Hall, K.R. and E. Ritchie. "A Study of Communication Behavior in an R&D Laboratory." *R&D Management* 5 (1975): 243-245.

15. Herner, Saul. "Information Gathering Habits of Workers in Pure and Applied Science." *Industrial and Engineering Chemistry* 46:1 (January 1954): 228-236.

16. Holmfeld, John D. *Communication Behavior of Scientists and Engineers*. Ph.D. Diss., Case Western Reserve University, 1970. UMI 70-25874.

17. Joenk, Rudy J. "Engineering Text for Engineers." Chapter 15 in *Technology of Text: Vol. II Principles for Structuring, Designing, and Displaying Text*, David H. Jonassen, ed. (Englewood Cliffs, NJ: Educational Technology Publications, 1985,) 346-369.

18. Kaufman, Harold G. *Factors Related to Use of Technical Information in Engineering Problem Solving*. (Brooklyn, NY: Polytechnic Institute of New York, January 1983.)

19. King, Donald W.; Jose-Marie Griffiths; Ellen A. Sweet; and Robert R.V. Wiederkehr. *A Study of the Value of Information and the Effect on Value of Intermediary Organizations, Timeliness of Services and Products, and Comprehen-*

siveness of EDB. (Rockville, MD: King Research, 1984.) (Available from NTIS, Springfield, VA DE82014250.)

20. Kremer, Jeannette M. *Information Flow Among Engineers in a Design Company*. Ph.D. Diss., University of Illinois at Urbana-Champaign, 1980. UMI 80-17965.

21. Krulee, G.K. and E.B. Nadler. "Studies of Education for Science and Engineering: Student Values and Curriculum Choice." *IRE Transactions on Engineering Management* 7:4 (June 1960): 157-158.

22. Landau, Ralph and Nathan Rosenberg (eds.) *The Positive Sum Strategy: Harnessing Technology for Economic Growth*. (Washington, DC: National Academy Press, 1986.)

23. O'Gara, Paul Webber. *Physical Location as a Determinant of Communication Probability Among R&D Engineers*. Master's Thesis, Massachusetts Institute of Technology, 1968.

24. Pinelli, Thomas E.; Myron Glassman; Walter E. Oliu; and Rebecca O. Barclay. *Technical Communications in Aeronautics: Results of an Exploratory Study*. TM-101534, Part 1, Washington, DC: National Aeronautics and Space Administration, February 1989. (Available from NTIS, Springfield, VA 89N26772.)

25. Price, Derek J. de Solla. "Is Technology Historically Independent of Science?" *Technology and Culture* 6:3 (Summer 1965): 553-578.

26. Price, Derek J. de Solla. *Science Since Babylon*. (New Haven: Yale Press, 1961.)

27. Ritti, R. Richard. *The Engineer in the Industrial Corporation*. (NY: Columbia University Press, 1971.)

28. Rosenberg, Victor. "Factors Affecting the Preferences of Industrial Personnel for Information Gathering Methods." *Information Storage and Retrieval* 3 (July 1967): 119-127.

29. Rosenbloom, Richard S. and Francis W. Wolek. *Technology and Information Transfer: A Survey of Practice in Industrial Organizations*. (Boston: Harvard University, 1970.)

30. Rothwell, R. and A.B. Robertson. "The Role of Communications in Technological Innovation." *Research Policy* 2 (1973): 204-225.

31. Rubenstein, Albert H; Richard T. Barth; and Charles F. Douds. "Ways to Improve Communications Between R&D Groups." *Research Management* 14:6 (November 1971): 49-59.

32. Schmookler, Jacob. *Invention and Economic Growth*. (Cambridge, MA: Harvard University Press, 1966.)

33. Shuchman, Hedvah L. *Information Transfer in Engineering*. (Glastonbury, CT: The Futures Group, 1981.)

34. Smith, C.G. "Consultation and Decision Processes in a Research and Development Laboratory." *Administrative Science Quarterly* 15 (1970): 203-215.

35. Solomon, Trudy and Louis G. Tornatzky. "Rethinking the Federal Government's Role in *Technological Innovation.*" In *Technological Innovation:*

Strategies for a New Partnership, Denis O. Gray, Trudy Solomon, and William Hetzner, eds. (NY: North-Holland Publishing Company, 1986,) 41-53.

36. System Development Corporation. *A System Study of Abstracting and Indexing in the United States*. Technical Memorandum WD-394. (Falls Church, VA: System Development Corporation, 16 December 1966.) (Available from NTIS, Springfield, VA PB 174 249.)

37. Taylor, Robert S. *Value-Added Processes in Information Systems*. (Norwood, NJ: Ablex Press, 1986.)

38. Young, J.F. and L.C. Harriott. "The Changing Technical Life of Engineers." *Mechanical Engineering* 101:1 (January 1979): 20-24.

39. Zipf, Geo K. *Human Behavior and the Principle of Least Effort*. (Cambridge, MA: Addison-Wesley, 1949.)

Physics Information
and Scientific Communication:
Information Sources
and Communication Patterns

Robert S. Allen

SUMMARY. The communication patterns of physicists and all scientists are undergoing change due to introduction of new technology, especially electronic communication. Traditional information sources of physics, as well as developing electronic information sources are defined. Also examined are the effects overlap and replacement of traditional information sources by new sources have on the communication patterns of scientists and the scholarly communication cycle.

INTRODUCTION

Much of the information seeking and transference in physics, and in other scholarly fields as well, has changed a great deal since the introduction of computers into the researchers' daily lives. To gain a basic understanding of the changes that are occurring in the communication dynamics of the physical and other sciences, it is necessary to examine the entire spectrum of information flow within today's scientific community. Much of the previous work that examines the scholarly communication cycle was directed at science before the computer era. The extremely rapid development

Robert S. Allen is Physics/Earth and Atmospheric Sciences Librarian at Purdue University, Physics Library, Physics Building, West Lafayette, IN 47907. He received a BS in geology from Eastern Illinois University, an MS in geology from Southern Illinois University, and an MS in library and information science from the University of Illinois at Urbana-Champaign.

and acceptance of electronic communication by physicists has created the need to reexamine the roles of traditional information sources and their interplay with the roles being assumed by electronic information sources.

INFORMATION SOURCES: CHANNELS OF COMMUNICATION

The methods used to transfer and gain scientific information are varied. To better define the interactions and philosophical implications of these communication methods, the methods themselves will first be defined. Though some of the following modes of communication are basic to any human interaction, they are included simply to offer a wide perspective on the flow of scientific information.

Oral Communication

This communication mode has historically been a heavily used but seldom documented method for communicating information in physics.[1] The most basic form of oral communication is informal personal interaction. It is quite natural that scientists who spend much of their time conducting research will tend to speak with colleagues about that research. This type of communication takes place wherever the scientist is, be it in an office, hallway, restaurant, library, gymnasium, etc. As most of the formal restraints common to many other communication modes are lacking, this is an opportunity for physicists to let their creativity flow,[2] and to trade bits of scientific gossip.[3] Conventional telephone conversation is an example of oral communication overlapping with the electronic modes to be discussed later. Another type of oral communication is formal personal interaction. This is characterized by unwritten rules, and is usually much less interactive than its informal counterpart. Common in professional conferences, seminars and colloquia, there is often a leader who presents an idea and then entertains questions or rebuttals from listeners. This formal framework often opens channels to more informal relationships between scientists. It is interesting to note that when physicists in the United Kingdom were asked to choose from a number of reasons why they attend conferences, most listed keeping in touch with current developments, meeting people with similar interests, and hearing papers presented in their field as being most important, though the list given to choose from was rather limited.[4]

Written Communication

There are many methods of transferring scientific information that use the written word. The literature of physics has been covered well elsewhere,[1,5] but the following short definitions are important here to show the interrelationships exhibited by communication patterns. The first mode to be defined is informal written communication. Though not highly revered by today's scientists as a means of documenting research, scholars of scientific history have devoted much time to studying personal letters left behind by great scientists. Lacking the formality of rigid scientific review, such communication has shown to be important in creating an environment of expression conducive to the formulation of many important scientific theories. This has been plainly shown by the extensive examination of the letters of Albert Einstein. Science information is also communicated through the popular news. This type of media is also outside the scrutiny of scientific review, and is often used as a soapbox for scientific quackery and unsound research. Not to be overlooked in importance to scientists, the popular news is a powerful force as it can often influence the various entities that award funding to carry out serious research.

Formal written communication is the most accepted form of reporting on scientific research within the scholarly community, though there are factors that influence the degree of acceptance of different types of scholarly publications. Types of formal written communication in physics are the refereed journal, preprint, monograph, conference proceedings, technical report, dissertation or thesis, popular journal, newsletter and abstracting journal.

The most accepted type of publication within the scientific community is the refereed journal. Research articles published in such journals are closely scrutinized by scientific referees to ensure that acceptable scientific methodology was used, and appropriate conclusions were made from the research. There are three general categories of refereed journals within the physics community, these being the standard journal, letters journal and review journal. The standard journal consists of rather long articles that include extensive discussion of methodology and data. These articles are considered to represent very substantial pieces of research and often are the culmination of a major research project. The letters journals are often associated with a corresponding standard journal, but articles are kept to a minimum length by editorial policy and are intended to provide a method for communicating recent research results in a timely fashion. Letters journals often include comments by other researchers about previous articles. Review journals are intended to include articles that encompass a broad field of research and rely heavily on previously published literature. They may or may not include a critical viewpoint.

Another type of formal written communication within the physics community is the preprint. Due to the international cooperation common in many fields of physics, research centers around the world make it a practice to circulate preprints of articles submitted by their researchers. It is not necessarily certain that the preprints will be published, and the type of publication in which they might appear varies greatly. Some might appear in a highly respected research journal, while others exist only as an internal technical report of the research center. Some preprints never appear in any form other than a preprint. The goal of the preprint is to quickly disseminate research information, avoiding the long publication delay commonly associated with refereed journals.

The monograph is often similar to the review journal, as many are intended to be a broad survey of a research field. Examples of such review monographs are often found used as the text for graduate courses. Undergraduate texts also rely heavily on review, though an important goal of these is to elicit interaction with the reader by using exercises. Monographs in physics are often biographical, with the authors eulogizing famous physicists with personal recollections coupled with collections of previous papers.

Though it often appears as a monographic work or as a special issue of a journal, the conference proceedings is unique enough to warrant a category of its own. This type of publication is created from the papers presented at a conference or special school, so the nature of the information contained therein is often similar to formal oral communication from the leader's perspective. Conference proceedings are often seen to contain review papers, letters type papers, and some papers similar to articles that might be found in a standard journal. Such papers, though often invited by conference planners who might also serve as referees for journals, are not necessarily to be considered refereed at the time of the conference. Conference papers often suffer in comparison to journal articles due to rapid preparation and length limitations.[16] At their best, conference proceedings function to bring current research quickly to the attention of a wide array of readers who did not attend the conference. Unfortunately, due to their often delayed final publication, conference proceedings function mainly as a written record of the conference and serve less well the needs of researchers looking for reliable scientific documentation.

Technical reports are usually documents not intended for wide use and which are not refereed. It is common for funded research to be accompanied by a technical report, especially when a governmental agency is doing the funding. Usually these reports give the status or final results of a large project, and are sometimes followed by a journal article describing the results of the project.

Dissertations and theses are created to document the results of long research projects carried out by graduate students. Though these documents are important in their own right, the journal articles often published as a result of the same graduate research are more widely disseminated and have the authority given by the refereeing process.

Physics information is also communicated in popular journals. Publications in this category often take a journalistic approach to science, rather than a primary reporting of refereed research results. These might be informal magazines intended to be light reading about physics for physicists, or they might be interdisciplinary scientific magazines containing physics as part of their subject matter. Similar to the informal magazine is the newsletter. Newsletters are intended to serve as a current awareness device, though some can contain hard research articles. Often they consist of columns and advertisements, similar to a newspaper meant to be read by scientists.

An attempt to bring all of these publications into one easily searchable indexed source is the abstracting journal. The scope of these secondary publications is determined by the body of primary publications being abstracted.

Electronic Communication

Modern communication has been indelibly marked by electronic information transfer. Many of the communication methods mentioned previously are being overlapped and in part replaced by electronic communication. Scientific information is communicated by the mass media through television and radio. Conferences can be televised via satellite to those unable to attend, greatly increasing the circle of participation. Much of society's daily spoken communication takes place over the telephone, and it is possible to hold an interactive mini-conference over the phone. Most businesses and universities today use telefacsimile machines to transfer documents, avoiding the delay associated with traditional mail and messenger services. Scientists make even greater use of electronic communication through the computer.

The computer and telecommunications have opened up new methods of scientific communication. One important source of information in physics is the electronic database.[7] Serving as reference sources these databases might be textual, numeric, image oriented, or a combination of these. The databases known best to librarians are textual databases that contain bibliographic information. Access to on-line catalogs of libraries is often available publicly, while access to other bibliographic databases is available commercially. It is common for scientists to keep a number of personally created databases so that experimental data or lists of references can be easily retrieved.[8] Information from such personal databases might be shared with colleagues, but for practical purposes these are private. A research group also compiles databases from numeric data created during the course of an experiment to be used in-house by the group. Research groups also band together to share a commonly created database on local levels,[9] national levels[10] and internationally." The number of international collaborators involved in very large experiments can be in the hundreds, making the need for remote access to centralized databases an absolute necessity.[12] The fact that much of the data collection done in physics research is carried out at research centers that are geographically distant from the researcher has created the development of state-of-the-art electronic networking.[13] Along with file transfer of experimental data has come a heavy usage of computers and telecommunications for other types of information exchange by physicists.

The large-scale collaborative nature of physics research has necessitated an extensive use of electronic mail by physicists. Electronic mail can be subdivided into a number of categories based on the dynamics of the information transfer being conducted. When information is transferred on a one to one basis, the resulting exchange can resemble informal oral, informal written and telephone communication. This type of electronic communication is the most exclusive and maintains the highest level of privacy for personal use. When electronic mail is used to transfer information from one to many, the dynamics become less personal though exclusivity is maintained by using a select list of receivers of information. This type of information transference structure becomes a special interest group. Information is also shared electronically through the use of bulletin board systems. The dynamics of this information exchange are characterized by a centrally maintained bulletin board that can be accessed by any number of interested parties. Selectivity here is maintained by the coordinators of the bulletin board, and a general interest group is formed by members having access to the board and interest in the information available. When many users of electronic mail transfer and receive information interactively, a computer conference is created.[14] A computer conference may have an official leader or moderator like its oral counterpart, but there is often a time lag between input and response of participants.

Primary publications are also beginning to appear in electronic format. Electronic newsletters exist,[15] and though they are quite similar to electronic bulletin boards, there is usually a person or group of persons responsible for editorial duties. Many "electronic journals" are available that are merely the textual contents of standard written journals in electronic format,[16] but these lack graphics and equations. The technology exists to create a stand alone electronic journal that contains all the information and figures found in a *standard paper journal*. Such an electronic journal would have the benefits of keyword searching and the drawbacks of hardware dependency, but a market for such technology has not yet been well established within the scientific community.

COMMUNICATION PATTERNS

The history of mankind's communication shows that the way we transfer information changes with developing technology. Communication patterns change as new media develop and man adapts his communication style to fit the media. When a new media develops it first overlaps and then, depending on its success, replaces the outdated media.

When a new communication technology is successful enough to replace an older technology there is usually an immediate benefit to the communicators. This immediate benefit often "halos" the opinion of the users of information, causing the potential hazards of the new technology to be overlooked or ignored. An example of this phenomenon is the brittle book problem. When paper chemists developed a technology that allowed for mass production of books using acid paper, the immediate benefits to readers far outweighed any concerns for potential problems that might have been encountered in the future. It is interesting to speculate upon what hazards the adoption of electronic communication will have on the communication patterns of scholars in the near future.

Electronic communication has been very successful because it provides an international and instantaneous channel for information flow. At present, it is overlapping many of the more traditional forms of communication. Personal letters have been overlapped, and to a great extent replaced, by the telephone. Both personal letters and the telephone are being overlapped and replaced by electronic mail. The immediate benefits of electronic information exchange are changing the face of scholarly communication, but the abrupt switch to a new media might create some problems in the future.

As more traditional forms of communicating are replaced, it is natural to assume that traditional documents will also decrease in number and possibly cease to exist altogether. An example of this is the replacement of personal written letters by electronic mail. Though the immediacy of electronic mail is of great benefit to scientists today, what will be the documentation available to historians of science in the future? With rapid developments in computer technology, it is difficult to predict whether outdated electronic media will be readable to future researchers.[17] The immediate nature of electronic mail requires that a researcher outside of the circle of primary researchers must "eavesdrop" electronically to produce a personal record of the communication of others. The author of the book *Cold Fusion* often cites "computer networks" as information sources,[15] but the ephemeral nature of such communication makes it difficult if not impossible for others to produce the same documentation today or in the future. Electronic mail also presents problems of confidentiality.[18] Because the informal oral and written communication it replaces is cited as being important to the creative process, in part due to scientists' being less accountable for what they say and write informally, the questioned confidentiality of electronic mail might inhibit creativity.

Electronic mail with many users involved is currently overlapping scientific newsletters. Bulletin boards and electronic mailing lists can replace many of the functions of paper newsletters, though the current extent of this replacement is unknown. When newsletters require extensive editorial direction and graphics, they are better produced by publishers with staff dedicated to these tasks. A typical electronic newsletter might be started by an individual who has access to the electronic mailing addresses of people who share a common interest in some topic. If interest in the topic dwindles, or if the person who started the newsletter loses interest in the topic, then the newsletter will die. Such ad hoc publications are quite common. As they are not usually intended to be archived as permanent scientific documentation, extensive replacement of paper newsletters by electronic media is probable.

Much has been written about electronic journals replacing paper journals. For this to occur, successful electronic journals must fulfill the same scholarly functions and financial returns as paper journals.[19,20,21] These scholarly functions are peer review for quality assurance, academic recognition for promotion purposes, international availability, format suitable for archival collections, inclusion in indexing and abstracting services, and consistent convenient access. Such a stand-alone electronic journal does not currently exist, though acceptance of CD-ROM and forthcoming graphics technology by publishers and users may bring about such publications. The library world has witnessed the release of many secondary publications on CD-ROM, though established paper counterparts usually pre-exist the electronic form. Secondary publications and bibliographic databases have been popular specimens for conversion because historic runs of textual data were available in machine-readable form and because graphics and equations were not usually included. Upon scanning the "instructions for authors" of journals today, submission in electronic format is usually encouraged. Though such prompting is intended to enable the publisher to create a finished product more easily, it must be assumed that standardized electronic submittal policies would also make it easier to convert to true electronic publication. It will be interesting to see whether some functions of scholarly communication influence electronic publishing more than others. If the function of academic recognition becomes more important to scientists than the function of archival quality, then the potential use of electronic journals by future researchers will be greatly hampered. Financial considerations are also important to the success of an electronic journal, and it must be assumed that publishers directed by profit-making motives would be less likely to replace an established and profitable paper journal with an electronic journal. The United States Government is currently pursuing electronic publishing as a means of saving money, and it is likely that professional and scientific associations will follow suit.

The effects of electronic communication on conferences and the like are also noticeable. As more researchers begin to use computer conferencing, it must be assumed that some of the benefits gained by attending traditional conferences will also be available. From the viewpoint of one who selects physics literature as part of library collection development, it is interesting to note the high number of very obscure conferences that appear in print as proceedings. One must wonder whether proceedings from computer conferences will also be available in the future, and in what format. Recently a large-scale interdisciplinary scientific teleconference on cold fusion was conducted, and the proceedings have been released on VHS videocassette.[22] The fact that teleconferencing technology was used improved the ability of the remote participants to immediately hear papers as they were read, though the ease with which non-participants can retrieve or browse such recorded proceedings in the future might be hampered by this medium.

CONCLUSION

The recent changes in the dynamics of scientific communication have been driven by technological developments in electronic communication. Of major concern to the scientific community is the need for prudent use of instantaneous electronic media for communicating scientific information. As mentioned earlier, the popular media is an important vehicle for influencing those who approve funding for science research. When scientists do not adhere to the rules of peer review for validation of research methods and release information prematurely, the controversy that can erupt causes damage to the reputation of scientists and their discoveries. The enormous benefits of electronic media to scientists and publishers must not outweigh the concern for potential damage to the scholarly communication cycle. The traditional rules of scientific communication must be enforced by scientists, and their choice of channels for communicating research results must adhere to these rules. Where there is overlapping and replacing of media, there is also the potential for destroying important archival records needed by researchers of the future. Much of the responsibility for creating such an archive and providing access to it will fall upon science librarians.

REFERENCES CITED

1. National Research Council. *Physics in Perspective*. Vol. 2b. Washington: National Academy of Sciences; 1973. 1465 p.
2. Kasperson, Conrad J. Scientific Creativity: A Relationship with Information Channels. *Psychological Reports*, 42(3.1): 691-694, 1978 June.
3. Traweek, Sharon. *Beamtimes and Lifetimes*. Cambridge: Harvard University Press; 1988; 187 p.
4. Hensman, Sandy. *Conference Attendance Patterns Amongst Physicists*. London: ASLIB; 1977; 52 p.

5. Shaw, Dennis F. *Information Sources in Physics*, 2nd ed. London: Butterworth's; 1985; 456 p.

6. Rowley, J.C. The Conference Literature – Savory or Acrid? In: Zamora, Gloria; Adamson, Martha C., eds. *Proceedings of The Workshop On Conference Literature in Science and Technology May 1-3, 1980*. Marlton, NJ: Learned Information; 1981; 220 p.

7. Gault, F.D. The Physics Database. *Computer Physics Communications*. 33(1-3); 1984 Aug./Sept.

8. Moon, C. Computerized personal information systems for research scientists. *International Journal of Information Management*. 8(4): 265-273; 1988 Dec.

9. Hughes, J. G. et al. The Atomic and Molecular Database at Belfast and Daresbury. *Computer Physics Communications*. 33(1-3): 99-103; 1984 Aug./Sept.

10. Lide, David R., Jr. The National Standard Reference Data System of the United States. *Computer Physics Communications*. 33(1-3): 207-210; 1984 Aug./Sept.

11. Whalley, M.R. The Durham-RAL High Energy Physics Databases – Hepdata. *Computer Physics Communications*. 57(1-3): 536-537; 1989 Dec.

12. Navarro, A. Savoy. Use of a data base in the on-line environment of a large-scale High-Energy-Physics Experiment. *Computer Physics Communication*. 33(1-3): 173-195; 1984 Aug./Sept.

13. Fluckiger, Francois. Overview of HEP wide area networking: producer Perspective. *Computer Physics Communication*. 57(1-3): 183-187; 1989 Dec.

14. Smith, Richard C. Teaching Special Relativity Through a Computer Conference. *American Journal of Physics*. 26(2): 142-147; 1988, Feb.

15. Peat, F. David. *Cold Fusion*. Chicago: Contemporary Books; 1989; 188 p.

16. STN International. *STN Contents Guide*. Washington: American Chemical Society; 1990; 49 p.

17. Reynolds, Chris. Just for the Record. *New Scientist*. 126(1714): 89; 1990; April 28.

18. Reynolds, Chris. Private and Confidential. *New Scientist*. 127(1726): 58; 1990; July 21.

19. Haar, Otto ter. Scientific Journals to the Year 2000. In: Fjallbrant, Nancy, ed. *Proceedings of the 11th Meeting of the IATUL Oxford, England April 15-19, 1985*. Goteborg: IATUL; 1985; 190 p.

20. Acock, Basil; Heller, Stephen R.; Rawlins, Stephen L. An Electronic Journal for Sharing Data on Crop Growth. In: Glaeser, Phyllis S., ed. *Proceedings of the Eleventh International CODATA Conference, Karlsruhe, Federal Republic of Germany, 26-29 September 1988*. New York: Hemisphere; 1990; 372 p.

21. Rogers, Sharon J.; Hurt, Charlene, S. How Scholarly Communication should Work in the 21st Century. *College & Research Libraries*. 51(1): 5-8; 1990 Jan.

22. *Workshop on Cold Fusion Phenomena*. Philips, David C., producer. Santa Fe: Los Alamos National Laboratory; 1989; 13 videocassettes.

Geologists and Gray Literature: Access, Use, and Problems

Julie Bichteler

SUMMARY. Geoscientists use large quantities of gray literature in the form of national, state, and local publications from societies and government agencies; dissertations and theses; maps; field trip guidebooks; and newsletters. Gray literature provides unique information on local and regional geology, oil and gas, soil, ground water, and mineral resources and is often produced more rapidly than traditional sources. Problems of physical quality, access, bibliographic control, and acquisition arise from inadequate coverage in bibliographies and databases, producers' lack of knowledge and concern for user needs, poor service from vendors, and university library cataloging and interlibrary loan policies. Recent improvements achieved by the American Geological Institute, geological surveys, and professional societies are encouraging.

INTRODUCTION

"Gray literature," a term which originated with British librarians, refers to information resources which are not available through conventional channels. These resources are frequently characterized by limited distribution, poor bibliographic control, small press runs, and nonstandard formats, i.e., literature which is out of the mainstream of standard access and acquisition. Wood defines gray

Julie Bichteler is Professor, Graduate School of Library and Information Science, The University of Texas at Austin, Austin, TX 78712. She received a BS (chemistry), MLIS, and PhD (library and information science) at The University of Texas at Austin.

The author is grateful to friends and colleagues who shared their knowledge and opinions of the many aspects of gray literature: Susan Klimley, John Mulvihill, Michael Noga, Miriam Sheaves, Janice Sorensen, Dennis Trombatore, and Rosalind Walcott.

literature as "that material which is not available through normal bookselling channels," such as technical reports, dissertations, government documents, semi-published conference proceedings, and maps.[1]

Information professionals view gray literature with ambivalence. Typically, this material bypasses accepted procedures followed in the scholarly communication process. When produced in-house without benefit of the external refereeing process, gray literature lacks validation from peers and loses credibility. Since gray literature has been excluded from conventional channels of acquisition and processing, librarians often find it time-consuming and troublesome. Small press runs and limited distribution add to their frustration. Yet the volume of gray literature continues to increase dramatically in proportion to the more well-established, traditional trade publications, and librarians must deal with it on a daily basis.

Regardless of problems connected with gray literature, there is no doubt that it contains information which a large number of people find valuable and useful. Unfortunately, lack of access has caused many users to be unaware of material which would satisfy their information needs.

This paper describes selected types of gray literature in geology, emphasizing those areas which are particularly significant to the field and those in which new developments have occurred. Problems in access, bibliographic control, and acquisition demonstrate the accuracy of the term "gray" when applied to these information resources.

GRAY LITERATURE IN GEOLOGY

Because of the nature of their discipline, geologists require a variety of gray literature. Geology is geographically oriented, thus geologists use publications of state, regional, and local societies and government agencies in addition to information resources from national societies and federal agencies, which all scientists use. Maps are vital. And, given the regional emphasis of their research, geologists find dissertations and theses especially valuable, as they often contain the best description of a particular locale.

Further complicating information access in geoscience is the wide variety of formats required by geologists, more than in other scientific disciplines. Maps, aerial photographs, well logs, cores, rock and mineral specimens, guidebooks, and informal field reports all offer challenges and problems. Of particular interest are the ways in which users become aware of individual items. As one would expect, given the fugitive, nonstandard nature of gray literature, personal contacts and serendipity play major roles.

Geologic Field Trip Guidebooks

Geologic field trips occur through a variety of activities, ranging from university geology courses to international congresses. Whenever and wherever geologists get together, a field trip to visit local sites of geological interest seems in order. Field trip guidebooks, produced for such excursions, provide a road log with descriptions of stops, a map of the area, charts of stratigraphic columnar sections, and articles which describe the local or regional geology. Frequently, local experts, who may be the field trip leaders, write the accompanying text.

Guidebooks are usually issued in series by a society, state survey, or university. Societies which issue guidebooks include local and regional geological societies and national societies and their regional sections. Several organizations may share in publishing guidebooks for a conference and sponsorship for a particular series may vary over the years. To complicate the life of the acquisition librarian even further, guidebook producers do not accept standing orders. Guidebook formats vary enormously, from a few sheets stapled together to bound volumes with color plates, fold-out maps, and photographs. Plastic spiral bindings, odd-shaped volumes, narrow margins inappropriate for binding, and cheap paper create problems in preservation for some guidebooks.

Geologists often find that guidebooks are the best and most recent source of information on the geology of a very specific area, thus they are in high demand. Produced much more quickly than refereed publications, they are very current and offer highly focused information, for example, details of an outcrop. Librarians, however, as Walcott pointed out recently, find these "sneaky, fly-by-night, changecoat publications hard to identify, hard to acquire, hard to catalog and retrieve, and hard to preserve. They are a librarian's ultimate challenge."[2]

Geologists become aware of guidebooks through citations in the literature, recommendations by colleagues, and personal knowledge of a particular field trip. In fact, Haner found that 23 percent of the authors who cited guidebooks in seven geological journals in 1985 had been either the editor of a guidebook or the author of a guidebook article.[3] She also noted that 21.4 percent of the articles in the same group of journals contained at least one citation to a guidebook.

State Survey Open-File Reports

The U.S. Geological Survey (USGS) and the surveys of the states produce a variety of publications such as reports of investigations, memoirs, special papers, maps, information circulars, bibliographies and bulletins, educational materials, and computer software. Such publications provide detailed descriptions and interpretations of a variety of topics, for example, oil and gas fields, ground water data, geothermal energy, mineral resources, geology of specific areas, soil surveys, and seismic studies and earthquake recordings. Traditional published reports are carefully edited, have high-quality illustrations, and are attractively printed. This process is expensive and time-consuming. By the time the word processors, cartographers, editors, and printers are finished and the report is ready for distribution, a year or more may have elapsed.

In contrast, most open-file reports (OFRs) issued by surveys are "quick and dirty" with less editing, cheaper reproduction and illustrations, and rapid release. Although all of the publications of geological surveys fall into the category of gray literature, state OFRs are perhaps the ultimate example and represent a large body of important geological literature. In an excellent report on state OFRs, Manson and Haner comment, "The percentage of state geological reports issued as OFRs is growing because open filing allows a state geological survey to maximize the amount of geological information it releases."[4] They go on to point out that, since 1980, state surveys have released more OFRs than published reports. Simi-

larly, in the case of the USGS the open-file report has long constituted the largest body of literature issued by the Survey.[5]

Although state geological survey reports are frequently cited in
geological journal literature, only one percent of those reports are
OFRs; on the other hand, OFRs account for fifteen percent of all
USGS reports cited.[4] This difference can be explained by the far
superior bibliographic control of USGS OFRs which has been in
place for several years.

Geologists find OFRs valuable since they cover the same subject
matter as that described above for standard survey publications.
However, a geologist's being aware that a relevant state OFR exists
has largely been a matter of being involved in the work personally,
of knowing what's happening in a state survey through colleagues,
or just plain luck. In the present study, the author examined articles
in nine primary journals in 1989-90 which cited state OFRs. A total
of 27 OFRs were cited. For 20 of these citations the connection
between the authors of the citing article and the OFR was obvious.
For example, one or more of the authors might be affiliated with the
state survey or working in the same city as the survey (usually at a
university). Only seven revealed no clear connection with the survey. In telephone interviews authors in the latter category explained
how they had located the OFRs cited. A couple were consultants or
former employees of the survey they cited; some explained how the
OFR had been mentioned in informal conversations with colleagues; two knew the authors of the OFRs through former association and kept up with their work. Not one had located the cited OFR
in a bibliographic source!

Research Newsletters

Research newsletters contain short technical notes, bibliographies, announcements, lists of papers in press, research reports, and
news. These inexpensive, informal publications facilitate cooperation and rapid communication among scientists and are the "invisible colleges made visible."[6] They are not refereed (the editor serves
chiefly as compiler) and are seldom found in libraries. Frequently
consisting of a few pages and appearing irregularly, depending on
the rate of accumulation of contributions, they keep the specialist

informed in much the same way as does his or her personal network.

Geologists praise the newsletters in fields such as paleontology, where descriptions of new species reach the specialist long before the same species are formally reported in the literature. Newsletters support communication in specialist areas such as trace fossils, carbonates, karsts, and phosphorites.[7]

Maps

Geologic maps are crucial to many areas of geoscience. Interpretations of geologic history, water resources research, exploration and development of mineral and energy resources, and investigating natural hazards all require the use of geologic maps.

Concern for the future of geologic mapping and for the needs of map users led to a recent National Research Council study by the Committee on Geologic Mapping of the Board on Earth Science.[8] This survey of geoscientists revealed the heaviest use of geologic maps to be in planning, exploration, and development of resources; scientific research and engineering; and hazard mitigation. Overwhelmingly, the greatest need was for "large-scale geologic maps." Unfortunately, interest in mapping is declining at the federal level and in the geological community in general, as fewer field geologists enter the profession. Mapping is time-consuming and not especially rewarding in terms of promotion, salary, and prestige.

A partial answer to this dilemma is the current revolution in mapping resulting from digital geologic and geographic information systems (GIS). With the ability to acquire, manipulate, and display digital data characterized by geographic coordinates, geologists can obtain integrated, customized maps for specific needs. Selected thematic overlays with choice of color, scale, projection, etc., provide a solid picture of an area. State and national surveys use GIS techniques and automated cartography to publish maps which can be continually updated and issued on demand.[9] For example, the Kansas Geological Survey no longer publishes maps using traditional procedures but produces them from digital databases on a large-format, computer-driven color plotter.[10] Thus, the Survey does not have to maintain large inventories of stored maps; revisions to the databases allow frequent update, and the cost of plotting limited-edition, low-demand maps is much less. The large number of combinations available to customers allow for an amazing variety of possible products. This one-of-a-kind production has proven enormously popular and has raised interesting questions for bibliographic control. What records should be kept on these custom-produced maps? What is the distinction between published and unpublished and unpublished maps? How should maps be archived, numbered, and identified bibliographically? Which maps should be listed in the Survey's "Catalogue of Publications"?

Map librarians use a variety of sources to acquire maps. In general, most geologic maps are obtained from in-house sources (32 percent), federal agencies (31 percent), and state agencies (20 percent).[8] Many state survey maps are released as OFRs, thus poor OFR access also affects a large segment of local maps. The Western Association of Map Librarians (WAML) has indexed USGS maps and state survey maps by topographic quadrangle name for four states thusfar, a valuable contribution to map access and acquisition. In addition to direct purchase from agencies and societies that produce maps, map librarians use vendors such as Geoscience Resources or GeoCenter (Germany) whose GeoKatalog, Volume 2, has an excellent index and refers users to reports in which maps may be found. (The library may already own the report!) Depositories along with gift and exchange programs may supply out-of-print maps. New products such as GEOINDEX (Earth Sciences CD-ROM from OCLC) and the *Publications of the USGS* on CD-ROM have been described by O'Donnell and Derksen.[11]

Dissertations and Theses

Dissertations and theses are especially important to geologists, as they may be the only sources of regional information. However, as Walcott reports, they are "surprisingly difficult to identify bibliographically and are even more difficult to obtain."[12]

In a recent effort to obtain 500 geoscience dissertations submitted from 1981 to 1985, she encountered a number of problems. University Microfilms International (UMI) includes no abstracts in the *Dissertations Abstracts Online* database if the dissertations are not available from UMI. UMI prices are quite high, and 7.5 percent of the desired dissertations were not available at all from that agency. Vital parts of some dissertations were unusable—maps were reproduced in fragments; plates lacked detail; lack of color for plates and maps was frustrating. Although UMI sells high quality 35-mm slides of maps and illustrations, color is still lacking. And microforms are time consuming to use.

Of the 500 dissertations studied, 20 percent were at universities that do not loan their own dissertations. This policy unfortunately forces researchers to purchase dissertations in order to look at them. Furthermore, users also need access to illustrations, and interlibrary loan appears to be the only way to accomplish this.

BIBLIOGRAPHIC CONTROL

Improved bibliographic control is the solution to many of the problems encountered in dealing with gray literature. Librarians and library users need better access to this material, especially in-depth subject access, which has traditionally been poor or nonexistent. For example, a thesis bibliography or state survey catalog typically offers only author and title indexes, if that, or at best provides title keyword access. The most obscure journal article receives better treatment in printed or on-line abstracting or indexing services. Recently, however, some encouraging developments have occurred in bibliographic control.

Bibliographies and Databases

The American Geological Institute (AGI) produces GeoRef, the on-line database for geoscience containing some 1,590,000 citations as of October, 1990. Providing access to geology of North America from 1785 and the geology of the rest of the world from 1933, GeoRef covers more than 3,000 journals in 40 languages, as well as books, maps, reports, USGS publications, theses, and dissertations.[13] In August 1990 GeoRef was offered on CD-ROM through SilverPlatter. Subscribers will receive quarterly updates.

The AGI continues to improve access to gray literature in geology. In 1989 John Mulvihill, director of GeoRef, contacted all U.S.

state geologists requesting that they submit copies of their 1989 state OFRs to the USGS library in Reston so they could be included in GeoRef and other indexes.[4] At the present time all but one of the surveys has agreed to supply copies of their OFRs, and these are currently being added to GeoRef. AGI will contact state surveys annually to ensure their continued contributions. State geologists confirm that OFRs are assuming more and more importance in the states' publishing output; AGI's move to add these publications to GeoRef is a major service to the geological community. One can speculate that citing patterns of state OFRs will change radically as these titles begin to appear on GeoRef search results. Students using GeoRef CD-ROM, for example, will see references to state OFRs whereas previously that student would have had to consult numerous state catalogs or OFR lists in a hit-or-miss operation or have an appropriate personal contact.

Using references compiled by the Geoscience Information Society, AGI is adding several thousand citations for theses and dissertations to those already in GeoRef. This project requires indexing, standardizing names of institutions, eliminating duplicates, and submitting lists to academic geoscience libraries for corroboration. When the current effort is completed in 1991, AGI will provide access to some 60,000 U.S. and Canadian titles and will also make available a printed bibliography of dissertations and theses.

Another of AGI's welcome improvements in bibliographic control is the addition of extensive notes concerning the diskettes accompanying publications such as USGS OFRs or Water Resources Investigations. Diskettes may contain programs or databases, and AGI describes the content, system requirements for use, format, etc. For example, USGS OFR 90-0466 is entitled "GEONAMES — Geologic Names of the U.S. through 1988" and is available on sixteen diskettes with eleven pages of accompanying documentation.

Professional organizations of geoscience information professionals have been influential in improving bibliographic control. WAML's recent contribution to map indexing, noted earlier, is only one example of that organization's accomplishments. The Geoscience Information Society has published the *Union List of Geologic Field Trip Guidebooks for North America* for more than twenty

years with the fifth edition appearing in 1989.'⁴ This indispensable
work allows one to identify and locate guidebooks, many of which
are out-of-print. The Society has also produced a useful, one-page
"Guidelines for Authors and Publishers" of geologic field trip
guidebooks aimed at improving their format, bibliographic description, distribution, and preservation. Wider publicity for new guidebooks is needed to facilitate acquisition.

Local Access and Maintenance

Large university libraries, hard-pressed financially and searching
for ways to cut costs, often institute cataloging policies which are
especially detrimental to the retrieval of gray literature. Several academic geoscience librarians describe their institutions' large, "un-
controlled" backlogs which, while waiting years for OCLC or
RLIN "hits" (which may never appear) are virtually inaccessible.
If the major research libraries don't contribute cataloging records
for gray literature to the networks, who will?

At The University of Texas at Austin, for example, the "minimum cataloging" (MINCAT) collection may be accessed only by
title (plus keyword) and author—no subject or series entry. And
keyword title searching becomes much more difficult with foreign
language titles! MINCAT processing is, of course, greatly prefera-
ble to materials remaining indefinitely in the backlog, and the geology librarian may request full cataloging for special items, an op-
tion obviously limited by available resources for technical services.

The library defines an OCLC hit as a record which can be used
without modification, thus USGS records, among others, can't be
used as they don't include LC call numbers. In the Geology Library
ten to fifteen percent of the new material ends up in the MINCAT
section, arranged by accession number, making browsing infeasi-
ble. Nearly all of this material is gray literature. A cursory examination revealed, for example, numerous federal and state publications
(reports from state surveys, water resources boards, Department of
Energy); publications of foreign geological surveys and other foreign government agencies; foreign and domestic society publications; theses and dissertations from the USSR, France, Germany,
Austria, Bolivia, and U.S. institutions which don't supply records
to OCLC; technical reports from university research bureaus and

institutes; field trip guidebooks; contractor and consultant reports; and gifts such as the proceedings of the 1986 Devonian Congress in China donated to the library by a professor who attended. The librarian estimates that one-half of the incoming Russian material goes to MINCAT. University librarians' concern over lack of access has led to a proposed pilot project to tag or create minimum-level OCLC records for these materials, a significant improvement over present policy.

CONCLUSION

Responsibilities for improving the quality of gray literature, its bibliographic control, access, and acquisition must be assumed by all concerned: producers, societies, libraries, information professionals, and geologists. Producers, in addition to determining quality and format, can improve availability simply by initiating appropriate distribution. Bibliographic services such as GeoRef and selected major collections and depositories should all receive copies. Libraries must then do their part by analyzing and cataloging gray literature; it may turn out to be more critical to their users than material which is traditionally given full processing.

As an intermediary, the AGI has assumed a premier role in improving access to gray literature in geology. GeoRef is undoubtedly the most outstanding geoscience bibliographic resource in the world and is invaluable to the entire geoscience community. One wonders if geologists appreciate GeoRef sufficiently! AGI has been extraordinarily receptive to suggestions from GeoRef users and has increased coverage of gray literature to an impressive extent. AGI and GeoRef deserve greater support from AGI's member societies.

Societies need to take greater initiative in dealing with the problems of gray literature. Their members can be highly influential, as was shown a few years ago in the successful effort to improve access to USGS OFRs. The Geoscience Information Society's projects in the areas of theses and guidebooks have been extremely effective. And end users themselves must get involved in issues of gray literature, just as scientists in all disciplines are doing in the case of primary journal problems. The information professions as a whole should support efforts to improve access with such programs as the System for information on Grey Literature in Europe, established a few years ago, or our own National Technical Information Service.

Gray literature is here to stay. It is irresistibly economical, fast, and convenient to produce and has proven to be a necessity to geologists. The time has come for the information community to take it seriously.

REFERENCES

1. Wood, D. N. The collection, bibliographic control and accessibility of grey literature. *IFLA Journal* 10(3): 278-282; 1984 August.

2. Walcott, Rosalind. Guidebook problems from the librarian's point of view. Ansari, Mary B., ed. *Proceedings of the Geoscience Information Society*; 1989 November 6-9; St. Louis, MO. Geoscience Information Society, 1990: p. 185-192.

3. Haner, Barbara E. Guidebook citation patterns in the geologic journal literature: a comparison between 1985 and 1967. Ansari, Mary B., ed. *Proceedings of the Geoscience Information Society*; 1989 November 6-9; St. Louis, MO. Geoscience Information Society, 1990: p. 159-169.

4. Manson, Connie J.; Haner, Barbara E. Research reports vary in availability. *Geotimes* 34(6): 15-16; 1989 June.

5. Ward, Dederick C.; Wheeler, Marjorie W.; Bier, Robert A., Jr. *Geologic Reference Sources*. 2d ed. Metuchen, NJ: Scarescrow Press; 1981: p. vii.

6. Wyatt, H. V. The invisible made visible. *Nature* v. 329, no 0 6137, Sept. 24, 1987.

7. Bichteler, Julie; Ward, Dederick. Information seeking behavior of geoscientists. *Special Libraries* 80(3): 169-178; 1989 Summer.

8. Mankin, Charles J. Geologic mapping: will needs be met? *Geotimes* 33(11): 6-7; 1988 November.

9. Van Driel, N.; Davis, J. C. eds. *Digital Geologic and Geographic Information Systems*. Washington, D.C.: American Geophysical Union; 1989.

10. Buchanan, Rex; Steeples, Don. On-demand map publication. *Geotimes* 35(4): 19-21; 1990 April.

11. O'Donnell, Jim; Derksen, Charlotte R. M. CD-ROM and floppy-disk databases for the earth sciences. Ansari, Mary B., ed. *Proceedings of the Geoscience Information Society*; 1989 November 6-9; St. Louis, MO. Geoscience Information Society, 1990: p. 5-33.

12. Walcott, Rosalind. Where have all the dissertations gone? *Geotimes* 33(4):7; 1988 April.

13. American Geological Institute. GeoRef database to be available on CD-ROM. Press Release, May 25, 1990: 1 p.

14. Geoscience Information Society, ed. *Union List of Geologic Field Trip Guidebooks of North America*. 5th ed. Alexandria, VA: American Geological Institute; 1989.

The Cold Fusion Story:
A Case Study
Illustrating the Communication
and Information Seeking Behavior
of Scientists

Victoria Welborn

SUMMARY. On March 23, 1989 a research team from the University of Utah announced at a press conference that they had achieved cold fusion in the laboratory. The events that ensued offer an opportunity to re-examine the information seeking and communicating process of scientists, with particular reference to peer review and the role of the scientific journal.

BACKGROUND CHRONOLOGY

On March 23, 1989 Stanley Pons, a professor of chemistry at the University of Utah, and Martin Fleischmann from the University of Southampton, reported at a press conference that they had achieved cold fusion in the laboratory. On this day articles appeared in the *Financial Times of London* and the *Wall Street Journal.* By the end of the week it was also known that there was another research team, headed by Steven Jones at Brigham Young University with a similar claim. Both research teams had submitted papers to *Nature* and preprints of papers were going out over computer bulletin boards."

Victoria Welborn is currently Head of the Science Library at the University of California, Santa Cruz. Prior to this she was Head of the Biological Sciences Library at The Ohio State University. She earned her BA in Biology at Wake Forest University in North Carolina and her MLS at Kent State University in Ohio.

While there are some conflicting reports on chronology and dates, the following seems to be the general series of events through July 1989.

Pons/Fleischmann began collaborating on cold fusion in the late '70s. They financed their work personally for several years. In 1988 the team submitted a research proposal to DOE asking for funding to continue the research.[40,42] In 1986 Steven Jones from Brigham Young University began work with cold fusion in metals. In the fall of 1988 he put together his ideas for a paper reporting on his research. In September he was asked to review the grant proposal from Pons/Fleischmann. Jones saw a possible collaboration with mutual benefits and in December 1988 Jones contacted Pons/Fleischmann about working together.[30,40]

In February of 1989 Pons and Fleischmann traveled to Brigham Young to meet with Jones. Discussions on joint publication were held but no decisions were made. Pons/Fleischmann wanted to continue their research quietly for about 18 months, and asked Jones if he would postpone reporting his results.[40,42] Jones informed his colleagues from Utah that he had submitted an abstract for the American Physical Society conference to be held in May and he still planned to honor that commitment. He was willing, however, to cancel a colloquium scheduled two days from the meeting and one of his graduate students canceled a talk at a research conference.[40]

On March 2 the DOE grant for $322,000 was awarded to Pons/Fleischmann. On March 6 the Presidents of University of Utah and Brigham Young University met to discuss the issues of collaboration and publication. It was agreed that the two teams would meet at the airport at Salt Lake City on March 24 and mail their papers simultaneously.[30,40]

In early March the U.S. editor of the *Journal of Electroanalytical Chemistry and Interfacial Electrochemistry* called Pons on a personal matter. The conversation turned to cold fusion, and the editor said he could get an article through the peer process rapidly.[42] On March 11 Pons/Fleischmann submitted an article to the *Journal of Electroanalytical Chemistry and Interfacial Electrochemistry.*[30,42] During the week of March 13 the University of Utah filed a patent based on the work of the Pons/Fleischmann research team. On approximately March 21 the University of Utah, due to concerns on patent rights, decided to announce the results of the fusion research in a press conference to be held in two days.[30,40,42] By March 22 the paper submitted to the *Journal of Electroanalytical Chemistry and Interfacial Electrochemistry* had been through the peer review process, changes and modifications had been made, and the revised document was in the hands of the U.S. editor.[42] On March 23 the now famous cold fusion press conference was held with Pons/Fleischmann announcing their work and belief that they had achieved cold fusion in the laboratory. On the same day articles appeared in two financial newspapers, the *Wall Street Journal* and the *Financial Times of London.*

When Jones heard about the press conference he believed that the agreement had been broken and on March 23 he sent his paper, originally scheduled for simultaneous submission on March 24, to *Nature.* Pons/Fleischmann submitted their paper to *Nature* on March 24.[40,42] On April 3 the article authored by Pons/Fleischmann appeared in the *Journal of Electroanalytical Chemistry and Interfacial Electrochemistry* under the heading "Preliminary Notes."[4]

The American Chemical Society held a special forum on cold fusion at their annual conference on April 12. Over 7,000 chemists attended the session where Pons explained his procedures and results. The chemists' reactions to Pons and his claims were favorable and some were calling it the "Woodstock of chemistry."[44] The chemists were also delighted that if cold fusion had been attained it had been by a chemist and not a physicist. ACS President Clayton Callis is reported to have said, "Now it appears that chemists may have come to the rescue."[44,11] However, Harold Furth, Director of the Plasma Physics Laboratory at Princeton, and referred to by one reporter as the "token nuclear physicist at the ACS session,"[44] asked about whether a control experiment had been run with light (or ordinary) water. Pons' response was somewhat oblique, stating that he had seen fusion in light water.[44]

On April 17 Pons held another press conference where he responded to the questions raised by Furth at the ACS conference. Pons reported that there was an "unexpected" production of heat in ordinary water, but once again he gave no numerical, empirical results.[28]

The April 20 issue of *Nature* announced that the next issue would have an article on the cold fusion research of Steven Jones, but there would be no article from Pons/Fleischmann. While both teams had submitted articles, Pons and Fleischmann had decided not to respond to the reviewer's comments and questions. Therefore *Nature* would not be publishing their article. Jones, however, had chosen to make the suggested modifications.[24]

On April 26 the House Committee on Science, Space and Technology held hearings on cold fusion. At this hearing Pons and Fleischmann stated that they were "sure as sure can be"[13] that cold fusion worked. Jones, however, reportedly made it clear that, while he did believe in the reality of cold fusion he saw it only as interesting physics, not as a potential technology for energy production.[1,13]

On May 1 Jones attended the American Physical Society conference. While Jones was met with a " . . . polite but generally sceptical reception . . . " and his claims " . . . survived the evening mostly unscathed . . . "[13] the response of the physicists to the work of the absent Pons/Fleischmann was mostly one of derision. One session reportedly "took on the atmosphere of a hanging party lacking only its intended victims."[13] A physicists' remark that the results perceived by Pons/Fleischmann was a sign of "incompetence, perhaps delusion" resulted in loud applause.[13,39]

On July 12 the Federal Cold Fusion Panel, established by the U.S. Energy Department, met and concluded that there was no convincing evidence that cold fusion had been achieved and recommended that there be no special grants to set up special labs or research centers.[14]

REFEREED ARTICLE VS. THE PRESS

After the initial excitement over cold fusion was followed by the realization that there had not been a major breakthrough, there was much discussion about the role of refereed journal articles and the use of the press. There were some feelings that the need for refereed journals had been substantiated. Daniel Koshland Jr. stated with regard to cold fusion, "The first lesson is that the merit in the established scientific procedure of exposing ones's findings to peer review before publicizing results is reaffirmed."[8] Others felt that refereed journals had been proven obsolete. One librarian stated,

"Scholarly journals are obsolete as the primary vehicle for scholarly communication. The recent furor over 'cold fusion,' for example, developed entirely outside the scholarly journal process."[40]

There are several demands and requirements inherent in publishing in a peer review journal. The research is going to be reviewed by scientists who will analyze and judge the quality of the science and the methodology. According to John Maddox, it is quite difficult to get a refereed article published without a control group. He states, " . . . referees can be relied upon almost without question to draw attention to control experiments that should have been carried out . . . "[17] Also, it is expected that the article will contain a full methodology section. In the March 1990 *Nature* it was stated in relation to the research on cold fusion " . . . that the first duty of researchers claiming new discoveries is to make the details available for the scrutiny of others."[27] So the process of preparing research for publication in a refereed journal would generally include the use of controls (or an explanation as to why controls were not used) and an extensive methodology section giving other scientists the details needed to judge and replicate the experiment. The work of Pons/Fleischmann proved deficient in both of these areas.

A story that appears in the press would have different characteristics reflecting the requirements of that media. In short, the press holds primary that the story be of a certain magnitude of importance and that the information be of a timely nature. The work of Pons/Fleischmann met these expectations.

THE ROLE OF PEER REVIEW AMONG SCIENTISTS

Scientists need information from each other in two ways: one, they need to be aware of other research going on in their fields and other fields; and two, they also need feedback from objective peers on their research, methodology, and experimental design. In a case such as cold fusion, physicists will ask different questions than chemists, or different labs will see the same information differently. Scientists communicate not only to validate their work, but to receive needed feedback to continue working.

In the classical model of scientific communication, this interaction, or peer review, begins when preliminary research is first presented at university seminars, then in forums such as a symposia or conferences. The scientists attending these forums not only learn what research is being done by other scientists, they provide the objective feedback pointing out the potential questions and/or modifications that are needed to advance the research to the point of publishing an article in a refereed journal. Peer review then begins in the early stages of the research, and continues through and beyond the process of publication.

Jones' documented plan for communication about his fusion research before he knew of the work of Pons/Fleischmann followed this classical model. After deciding to prepare an article he planned a seminar on his work, and then submitted an abstract to present at the American Physical Society's conference. After discussions with Pons/Fleischmann, the plan did change and Jones submitted his article to *Nature* before his presentation at the American Physical Conference.

Pons/Fleischmann have been reported as stating that they would have preferred about 18 more months of time before they published.[40,42] After meeting with Jones they planned to publish in March before Jones presented his research at a conference in May. The University of Utah, due to concerns about patent rights, having "agonized" over the decision decided to hold a press conference concerning cold fusion before the articles were submitted.[40]

Therefore both research teams ultimately submitted a full article for publication in *Nature* before benefiting from the feedback normally received at seminars and conferences.

THE SCIENTIFIC COMMUNITY'S REACTION TO COLD FUSION

The immediate reaction of the scientific community was one of great excitement. Even though the information presented on methodology was somewhat sketchy there were reports of scientists all around the world working to replicate the results Pons/Fleischmann claimed they had accomplished in their laboratory.[23,29] Scientists were willing to find the information they needed through any means available. They used press articles, telephones, electronic bulletin boards, and fax machines. The rush for information was such that one scientist was reported to state that without a fax machine you were nowhere.[3] In general it seems that scientists were less concerned with the source of their information as long as there was a mechanism to get it. While a peer review article would have been the most conventional method, they were willing to use other methods.

As scientists asked more and more questions about the methodology and results of Pons/Fleischmann, the frustration levels increased. It appeared that Pons/Fleischmann were withholding information.[3,12,27,35] There was reported speculation on the reasons for this, the advice of lawyers, the desire of Pons/Fleischmann, or the fact that they did not have the answers.[35] By March 1990, a year after the initial press conference of Pons/Fleischmann, David Lindley stated "During the past year, the original claims of Pons and Fleischmann have diminished, the experimental evidence has been subtracted from not added to."[12] Had Pons/Fleischmann gone the conventional route many of the problems with the methodology would have been pointed out by colleagues in seminars, workshops, and informal conversations.

Pons/Fleischmann did author the article which appeared in the *Journal of Electroanalytical Chemistry and Interfacial Electrochemistry* which had undergone peer review. However, the journal received criticism for publishing the article in its form. The journal's response to the criticism was that they felt the information was important enough to publish and they did publish the piece under the heading "Preliminary Note," indicating a lack of information.[42] Both research teams had submitted articles to *Nature*. After the initial peer review, Jones made the necessary modifications so that his article was published in the journal. Pons/Fleischmann, however, declined to make the modifications and therefore *Nature* chose not to publish the article.[24]

CONCLUSION

The cold fusion story has done much to clarify the role of the refereed journal and peer review in science communication. On the one hand Pons/Fleischmann and the University of Utah felt a need to publicly announce their work to ensure patent rights. Believing that they had found the means for a significant production of energy with global implications, they also believed they did not have the time to wait to publish in a refereed journal. Because of this they went to the press with their information before their research results and methodology had undergone any kind of peer review.

The press did provide an avenue and process that offered a quick release of information. But the in-depth details needed to replicate the experiments of Pons/Fleischmann, not expected in a press article, were never made available to the scientific community through any means. Inevitably the cold fusion work of Pons/Fleischmann did go through peer review, but it was the last step of the process, instead of one of the first. It was done by the world community of chemists and physicists, not by selected representatives.

The refereed article or forum offers implicit guarantees to scientists receiving the information. The referees act somewhat like a screening committee. Articles are reviewed for importance and methodology. Therefore, in theory, to read an article in a peer reviewed journal is to learn of research which has been reviewed by scientists assuring that a certain level of methodology and design has been conducted, and that this research is a significant contribution to the literature and to the science. While the process of submission, review, amendment, and resubmission is time consuming to the producer it may ultimately be one of the most efficient ways for the scientific community to receive information.

Pons/Fleischmann did have a perceived need for a rapid release of information that was best handled by the press. However, the fact that the cold fusion furor "developed entirely outside the scholarly journal process" was more a reflection of the state of Pons/Fleischmann's research than the nature of refereed journals.

REFERENCES

1. Crawford, Mark. Utah looks to congress for cold fusion cash. *Science.* 244(2): 522-523; 1989 May 5.

2. Crawford, Mark. Cold fusion: Is it hot enough to make power? *Science.* 244(1): 423; 1989 April 28.

3. Crosariol, Beppi. "Cold fusion" and the Media. *Queen's Quarterly.* 96(4): 815-822; 1989 Winter.

4. Fleischmann, Martin; Pons, Stanley. Electrochemically induced nuclear fusion of deuterium. *Journal of Electroanalytical Chemistry and Interfacial Electrochemistry*. 261: 301-308; 1989.

5. Garwin, Richard L. Consensus on cold fusion still elusive. *Nature*. 338: 616-617; 1989 April 20.

6. Hively, William. Cold fusion confirmed. *American Scientist*. 77: 327; 1989 July-August.

7. Jones, S. E.; Palmer, E. P; Czirr, J. B; et al. Observation of cold nuclear fusion in condensed matter. *Nature*. 388: 737-740; 1989 April 27.

8. Koshland, Daniel E. Jr. The confusion profusion. *Science*. 244(2): 753; 1989 May 19.

9. Levi, Barbara G. Doubts grow as many attempts at cold fusion fail. *Physics Today*. 42: 17-19; 1989 June.

10. Lindley, David. Cold fusion gathering is incentive to collaboration. *Nature*. 339: 325; 1989 June 1.

11. Lindley, David. Double blow for cold nuclear fusion. *Nature*. 339: 567; 1989 June 22.

12. Lindley, David. The embarrassment of cold fusion. *Nature*. 344: 375-376; 1990 March 29.

13. Lindley, David. More than scepticism. *Nature*. 339: 5; 1989 May 4.

14. Lindley, David. No new money from US government? *Nature*. 340: 174; 1989 July 20.

15. Lindley, David. Noncommittal outcome. *Nature*. 341: 679; 1989 October 26.

16. Lindley, David. Still no certainty. *Nature*. 339: 85; 1989 May 11.

17. Maddox, John. Can journals influence science? *Nature* 339: 657; 1989 June 29.

18. Maddox, John. End of cold fusion in sight. *Nature*. 340: 15; 1989 July 6.

19. Maddox, John. Publishing without being damned. *Nature*. 343: 113; 1990 January 11.

20. Maddox, John. What to say about cold fusion. *Nature*. 338: 701; 1989 April 27.

21. Maddox, John. Where next with peer-review? *Nature*. 339: 11; 1989 May 4.

22. Cold (con)fusion. *Nature*. 338: 361-362; 1989 March 30.

23. Cold fusion causes frenzy but lacks confirmation. *Nature*. 338: 447; 1989 April 6.

24. Cold fusion in print. *Nature*. 338: 604; 1989 April 20.

25. Cold results from Utah. *Nature*. 338: 364; 1989 March 30.

26. Disorderly publication. *Nature*. 338: 527-528; 1989 April 13.

27. Farewell (not fond) to cold fusion. *Nature*. 344: 365; 1990 March 29.

28. Hopes for nuclear fusion continue to turn cool. *Nature*. 338: 691; 1989 April 27.

29. Hot-footed towards cold fusion. Nature. 338: 535; 1989 April 13.

30. Prospect of achieving cold fusion tantalizes. *Nature*. 338: 529; 1989 April 13.

31. Scientific look at cold fusion inconclusive. *Nature*. 338: 605; 1989 April 20.

32. The cold fusion controversy: sizzle or fizzle. *Plating and Surface Finishing*. 76(9):40-41; 1989 Sept.

33. Pool, Robert. Bulls outpace bears — for now! *Science*. 244(1): 421; 1989 April 28.

34. Pool, Robert. Cold fusion: Bait and switch? *Science*. 244(2): 774; 1989 May 14.

35. Pool, Robert. Cold fusion: End of act I. *Science*. 244(3): 1049-1049; 1989 June 2.

36. Pool, Robert. Cold fusion: Smoke, little light. *Science*. 246(2): 879; 1989 Nov. 17.

37. Pool, Robert. Cold fusion still in state of confusion. *Science*. 245: 256; 1989 July 21.

38. Pool, Robert. Confirmations heat up cold fusion prospects. *Science* 294(1): 143-144; 1989 April 14.

39. Pool, Robert; Heppenheimer, T.A. Electrochemists fail to heat up cold fusion. *Science*. 244(2): 647; 1989 May 12.

40. Pool, Robert. Fusion followup: Confusion Abounds. *Science*. 244(1): 27-29; 1989 April 2.

41. Pool, Robert. Fusion theories pro and con. *Science*. 244(1):285; 1989 April 21.

42. Pool, Robert. How cold fusion happened — Twice. *Science*. 244(1): 420-423; 1989 April 28.

43. Pool, Robert. In hot water over cold fusion. *Science*. 246(3): 1384; 1989 Dec. 15.

44. Pool, Robert. Skepticism grows over cold fusion. *Science*. 244(1): 284-285; 1989 April 21.

45. Pool, Robert. Some companies keep a foot in the door. *Science*. 245: 256; 1989 July 21.

46. Pool, Robert. Teleer, Chu "Boost" cold fusion. *Science*. 246:449; 1989 Oct. 27.

47. Pool, Robert. "Utah effect" strikes again? *Science*. 244(1): 420; 1989 April 28.

48. Pool, Robert. Will new evidence support cold fusion? *Science*. 246: 206; 1989 Oct. 13.

49. Rogers, Sharon J.; Hurt, Charlene S. How scholarly communication should work in the 21st. century. *College and Research Libraries* 1990 January.

50. Utah scientist: No cold fusion. *Science*. 248: 36; 1990 April 6.

51. Swinbanks, David. Efforts abandoned in Japan. *Nature*. 339: 167; 1989 May 18.

Informal Communication Among Scientists and Engineers: A Review of the Literature

Jean Poland

SUMMARY. The literature dealing with informal communication behavior among scientists and engineers is reviewed. The effects new communications technology may have on that behavior are considered, along with implications for librarians.

INTRODUCTION

The study of information transfer among scientists and engineers is basic to library and information science. In the 1960's and 1970's research dealing with interaction in scientific and technical communities flourished. Professional societies and federal agencies funded long-term studies that led to modelling the development and transfer of knowledge in science and technology. Over time, policy recommendations were formulated regarding national and international information transfer.

Herbert Menzel's early description of empirical studies of scientific communication continues to be appropriate:

> When approached from the point of view of the individual scientist, these are studies of "scientists' communication behavior." When approached from the point of view of any communication medium, they are "use studies." When approached from the point of view of the scientific communication system, they are studies in the flow of information among scientists.[1]

Jean Poland is Assistant Engineering Librarian, Purdue University, West Lafayette, Indiana 47907. She holds a BA from the University of New Hampshire and an MS in Library Science from the University of Illinois.

This review will concentrate on the first group, that of scientists' and engineers' communication behavior, with particular emphasis on informal communication. Concepts fundamental to the study of informal scientific and technical communication will be summarized. A large body of literature on the subject has already been compiled in bibliographies and reviews. These will be listed and some reports that have appeared in the past decade will be summarized. The effect new communications technologies may have on informal communication will be considered, followed by some implications for librarians.

The characteristics of informal scientific and technical communication were well summarized by Garvey and Griffith in 1971.[2] They wrote that informal communication is relatively free of filtering, is unfinished in nature, sometimes redundant, may function as the beginning of a feedback loop, and is not stored in any permanent way. The flow is initiated and to some degree controlled by the researcher. Writing on the differences between formal and informal interaction, Pfaffenberger compared the tacit knowledge involved in informal communication with the textual knowledge of written information.[3] Trial and error experimentation and the emphasis that technological thinking places on visualization are two factors that show the limitations of published literature as a means of scientific and technical communication. Informal networks are ideal for the expression of hypothetical thinking.

Much informal communication consists of conversations and comments on drafts of papers or reports. It occurs before work is published or in some other way permanently recorded in the literature but it does not always lead to publication especially among people in industrial settings where the end product is often a design.

Because informal communication was perceived to be inefficient and exclusive, attempts were made to "formalize" the process. Wolek and Griffith discussed some of these efforts in their outline of a systems context for policy on informal communication.[4]

FUNDAMENTAL CONCEPTS

The concepts that are now considered basic to our understanding of scientific and technical communication were developed over decades. Results tend to indicate that a sense of group cohesiveness develops among people working on the same or related projects often leading to limited interaction with nonmembers of the group. Interaction outside the group appears to increase with the degree of uncertainty members have regarding their own knowledge of the subject.

Garvey and Griffith's work in psychology, under the auspices of the American Psychological Association (APA), was one of the first comprehensive studies of scientific information exchange.[5] Griffith and Miller summarized the findings of the APA project regarding informal communication.[6] They reported a relationship between the coherence of a group and the degree to which that group perceived its work to be important to the field. Recognizable leaders often emerged in these groups.

Derek de Solla Price articulated the notion of the invisible college in his report of the activities of a National Institutes of Health sponsored scientific exchange group.[7] Diana Crane made a major contribution to the study of invisible colleges with her examination of interaction within and between scientific communities.[8] She based her research on the diffusion of agricultural innovations among rural sociologists, a relatively new field which was considered to be a research front at the time. Although much of her work focussed on formal communication, she described informal communication as particularly important for transferring new information. In her study of the social organization of research areas, she determined that groups of "collaborators" existed and that communication took place within and across these groups. As the interests of a group approached new or unknown research, informal communication increased.

In 1974 Thomas Allen published the results of a decade of research.[9] He integrated several related studies into a theory of the communication process in technology. Allen developed the concept of technological gatekeepers, people who seem to have most contact with other members of the group and who function as informal information disseminators. He also reported that the probability of weekly communication exists in any significant way only for people whose offices are within thirty meters of each other. Correcting for organizational bonds, Allen found that while the probability of communication increased with stronger organizational bonds, thirty meters remained the point at which there was no significant change in the probability that communication would occur.

Another part of Allen's study alluded to the informal written literature comprised primarily of internal and external technical reports. He noted that such material was most often borrowed from colleagues or kept in a personal file. It was seldom borrowed from a library.

Katz and Allen looked at the effect of long-term membership on project teams.[10] They found that when research and development groups maintained the same membership for more than several years, communication decreased among the members and between members and external colleagues. This was reflected by the attitude that the team knew all there was to know about the project and no one outside the group could contribute significant information. If a product was "not invented here" it was not worth knowing about.

BIBLIOGRAPHIES AND REVIEWS

In 1960, Columbia University sponsored Menzel's analysis of empirical studies of scientific communication. Data from twenty-six reports were organized in a complex series of tables comparing the various studies, their approaches, and their results.[11] Six years later, A. Stephanie Barber looked at some of the same studies.[12] Her unique contribution was a five-page citation index to the research under review. In his 1965 review, Fishenden noted that design details are not always recorded but often transmitted orally suggesting that engineers would be particularly appropriate subjects for studies of informal scientific and technical communication.[13] In the same journal issue, Barnes examined the methodology of several surveys.[14] Each of these early works noted the disparity in methodology among the various studies and suggested that common ground be explored and used as a base for further research.

In 1967, Rutger's Graduate School of Library Services published a massive classified bibliography that covered, for the period 1955-1965, "all aspects of generation, acquisition, processing, storage, retrieval, and use of information."[15] Over one hundred annotated entries are included under "use in science and technology."

From its inception, *Annual Review of Information Science and Technology (ARIST)* became the source of record for information about scientific information. Within a few years, the original title, "Information Needs and Uses in Science and Technology," was broadened to "Information Needs and Uses" but, in most instances, the emphasis remained on scientific and technical information. The consistency with which the chapters appeared in the first seven volumes, and the randomness of publication of these chapters in the following years, reflects to some degree the lessening of interest in the subject of scientific communication. Each author of an *ARIST* chapter on information needs and uses had established himself or herself as an important contributor to the field.

In the first volume of *ARIST*, Menzel presented a critical review of studies from 1963 through 1965 particularly praising the methodology of the APA project.[16] In this chapter, he classified studies of informal communication under a heading that included "inefficient" communication activities.

The following year, Saul and Mary Herner took a less critical and more inclusive approach to their review, but they too felt that research techniques were not as good as they might be.[17]

In *ARIST* Volume 3, Paisley reported a "bumper crop" of studies during late 1966 and 1967.[18] He discussed the merging of information science and the behavioral sciences in the study of information needs and uses. Paisley saw scientists at the center of concentric circles, the larger one representing the invisible college, and the smaller circle representing the workgroup.

In the following annual volume, Allen built on Paisley's context structure.[19] He noted that the increase in the number of studies of information transfer among social and behavioral scientists signaled the legitimization of the field and reflected the discomfort behavioral scientists felt when they studied physical scientists and engineers.

In a 1970 *Special Libraries* article, Janice Ladendorf succinctly synthesized a decade's work.[20] She presented a summary of "two-stage information flow" where informal information passes through people who function as gatekeepers in both technology and science. *ARIST* began the new decade with Ben-Ami Lipetz's review that returned to an emphasis on scientific and technical information.[21]

Crane's 1971 *ARIST* article divided information need and use studies into those in the basic sciences and those in technology.[22] A third section looked at international aspects.

A 1972 bibliography by Jacqueline Hills on primary communications included a section on information transfer that emphasized international aspects of information transfer.[23] The internationalization of information use studies was further highlighted by Lin and Garvey.[24]

A. J. Meadow's comprehensive review of communication in science demonstrates the complexity the subject had reached.[25] Although he emphasized formal communication channels, Meadows presented a concise summary of research on informal communication.

In *ARIST* Volume 9, John Martyn noted that use studies were employing social science methods and systems approaches.[26] In the next *ARIST* chapter on information needs and uses, in 1978, Susan Crawford reported a growth rate of more than 30 studies a year for the period 1975 through 1977.[27] Studies in physical sciences and technology constituted only one quarter of those. The post-1978 literature was reported by Brenda Dervin and Michael Nilan in *ARIST* Volume 21.[28] They found no significant decline in the number of published studies. In much the same way that Menzel had sixteen years earlier, they called for a new approach that would emphasize conceptual growth.

AN UPDATE

Bibliographies and literature reviews of information transfer have of necessity been selective. One publication medium that has been relatively ignored is dissertations produced by doctoral candidates at schools of library and information science. These studies are frequently as useful for their literature reviews as they are for their results.

Jeannette Kremer described the information transfer processes in an Illinois engineering firm.[29] She found that all respondents, regardless of years of experience, preferred using internal channels for information. She also saw a discrepancy between the way engineers reported their behavior and the way they really functioned. While engineers reported that they used colleagues as one of three major information sources, in reality they were reluctant to ask in-house colleagues for help preferring to use books and manuals, which they considered more authoritative.

Edet Efiong Nkereuwem analyzed information use by scientists and engineers in the Nigerian petroleum industry.[30] Comparing how engineers and scientists used informal communication, he found that engineers depended on discussions with colleagues both within departments and in other organizations while scientists depended to a greater extent on workshops and discussion with colleagues in universities.

The most pragmatic reason for studying information transfer is the possibility of incorrect application of information. The results of too much dependence on informal sources can be disastrous if the informal group does not generate a correct answer. Patterson and Farrant studied members of the New Zealand building industry for that reason.[31] They report that in fact most decisions are based on previous personal experience, information from colleagues, and written material, in that order.

William Lacy and Lawrence Busch studied information exchange among agricultural scientists from different departments.[32] They found that most interaction occurred within departments, but only on a weekly basis. Communication with scientists outside the department and with people outside the agricultural sciences arena was even less frequent. Lacy and Busch also looked at which sources influenced agricultural scientists in deciding research methods. Colleagues at other institutions were influential slightly more frequently than were department colleagues. The authors hypothesize that information exchange with nondepartmental colleagues occurs at professional meetings.

Hedvah Shuchman's survey of engineers working in industry in six major disciplines has become a landmark study.[33] Results indicate that, in general, engineers begin with their own store of information, then talk first with colleagues, then with supervisors. If they feel further need of information they turn to technical reports, gatekeepers in the firm, and finally to various formal sources. Shuchman found that half of the engineers surveyed had weekly telephone contact with engineers outside their own workgroup.

As part of the National Engineering Utilization Survey Jones, LeBold, and Pernicka surveyed engineers working in industrial settings.[34] Respondents were asked to rank a list of sources of information used in the previous year for "keeping up-to-date" and for solving problems on current projects. Informal contact with co-workers, personal store of information, and catalogs and manufacturers' literature were reported as used most frequently for both keeping up-to-date and in current projects. Internal technical reports and handbooks followed as most consulted for current projects while published technical articles, external technical reports, and internal technical reports were reported to be used more for keeping up-to-date. The results are further broken down by area of engineering specialization. Informal contact with co-workers, catalogs and manufacturers' literature, and published technical articles were ranked as equally used by mining engineers while chemical engineers ranked published technical articles after informal contact with co-workers and personal store of information.

The aerospace industry has been very active in researching information transfer. D. A. Raitt surveyed scientists and engineers in three national and three international aerospace organizations.[35] He found that colleagues were the main source for all kinds of information. He found that people preferred face-to-face discussion in order to be able to read facial features for further feedback.

Thomas Pinelli and his colleagues at NASA's Langley Research Center have been studying technical communications in aeronautics for several years.[36,37,38] Their findings regarding informal communications do not differ significantly from Shuchman's. The first four sources reported by respondents are personal knowledge, informal discussion with colleagues, discussion with experts within the organization and discussions with supervisors. High use of electronic communications technologies such as FAX or TELEX, teleconferencing, and electronic mail were also reported.

NEW DIRECTIONS

Writing in 1970, Ladendorf was able to divide a decade of research into two groups of writings: those dealing with "written communication" and those dealing with "oral communication."[39] New communication media are now almost routinely used in the scientific and technical community. Electronic mail, electronic conferencing, and electronic publishing are changing the process of information seeking. Just as searching a CD-ROM bibliographic data base involves the same intellectual processes but is different from searching a printed index, electronic communication of scientific and technical information may result in adjustments to the way informal information transfer takes place. The most obvious change is elimination of the distance factor.

Wilfrid Lancaster forecast the paperless society a decade ago.[41] In his vision computers would be used to send, receive, and store communications. His predictions seemed outrageous to some, but the kinds of electronic media he discussed are now widely used. Ivars Peterson addresses the effects electronic mail and telefacsimile have had on the way scientists work today.[40] He describes the various computer networks and the ways in which researchers are using those networks: reading electronic mail, transmitting computer files, collecting data from remote instruments. To illustrate the new electronic community, Peterson reports on the events surrounding a mathematical breakthrough using computer collaboration.

IMPLICATIONS FOR LIBRARIANS

There appears to be no study of information seeking behavior that shows libraries as the first place researchers look for information. Libraries are in effect part of the formal spectrum of information transfer, and as such, among the last places scientists and engineers look for information.

As he describes the characteristics of the organizational gatekeeper, Allen writes: "librarians and other purveyors of information might well be able to perform their functions better if they were to know the identity of the gatekeepers in their organization."[42] By identifying and interacting with the gatekeepers in his or her organization, the librarian can be more involved in the informal information communication network.

Many research and development organizations with geographically scattered offices use international networks to maintain contact among divisions. Giuliana Lavendel describes how librarians at Xerox use electronic communication to deliver information services.[43] Librarians in all segments of the economy will need to make greater use of the networks available to them, not only to assist their patrons, but also to enhance their own informal networks.

Garvey gives detailed suggestions about how librarians might study the information seeking activities of their particular patron groups. His closing sentences provide the most important advice:

. . . If genuine collaboration is to come about, then it will have to be a matter of librarians making the effort to study and understand these aspects of scientific communication so they can communicate with scientists on their own grounds. When scientists are convinced that librarians understand why communication is the essence of science, then librarians will find that they will have some enthusiastic, collaborating scientists on their hands."

NOTES

1. Columbia University. Bureau of Applied Social Research. *Review of Studies in the Flow of Information Among Scientists.* 2 volumes. January 1960: p. 1.

2. Garvey, William D.; Griffith, Belver C. Scientific communication: its role in the conduct of research and creation of knowledge. In: Griffith, Belver C. ed., *Key Papers in Information Science.* White Plains, New York: Knowledge Industry Publications; 1980. p. 38-51. (Reprinted from *American Psychologist.* 26(4); 1971 April.)

3. Pfaffenberger, Bryan. *Democratizing Information: Online Databases and the Rise of End-User Searching.* Boston: G. K. Hall; 1990. 191 p.

4. Wolek, Francis W.; Griffith, Belver C. Policy and informal communication in applied science and technology. In: Griffith, Belver C. ed., *Key Papers in Information Science.* White Plains, New York: Knowledge Industry Publications; p. 113-124. (Reprinted from Science Studies. 4; 1974.)

5. American Psychological Association. *Reports of the American Psychological Association's Project on Scientific Information Exchange in Psychology.* Washington, D.C. Volume 1, Overview Report and Reports no. 1-9, December 1963; Volume 2, Reports no. 10-15, December 1965.

6. Griffith, Belver C.; Miller, A. James. Networks of informal communication among scientifically productive scientists. In: Nelson, Carnot E.; Pollock, Donald K. eds. *Communication Among Scientists and Engineers.* Lexington, Massachusetts: D. C. Heath and Company; 1970. p. 125-140.

7. Price, Derek J. de Solla; Beaver, Donald deB. Collaboration in an invisible college. *American Psychologist.* 21; 1101-1117, 1966.

8. Crane, Diana. *Invisible Colleges: Diffusion of Knowledge in Scientific Communities.* Chicago: University of Chicago Press; 1972. 213 p.

9. Allen, Thomas J. *Managing the Flow of Technology.* Cambridge: MIT Press, 1977. 320 p.

10. Katz, Ralph; Allen, Thomas J. Investigating the not invented here (NIH) syndrome: a look at the performance, tenure, and communication patterns of 50 R&D project groups. *R&D Management.* 12(1): 7-19; 1982.

11. Columbia University. Bureau of Applied Social Research. *Review of Studies in the Flow of Information Among Scientists. op. cit.*

12. Barber, A. Stephanie. A critical review of the surveys of scientists' use of libraries. In: Saunders, W. L. ed., *The Provision and Use of Library and Documentation Services.* Oxford: Pergamon Press; 1966; p. 145-179.

13. Fishenden, R. M. Information use studies part 1 — past results and future needs. *Journal of Documentation.* 21(3): 163-168; 1965 Sept.

14. Barnes, R. C. M. Information use studies part 2 — comparison of some recent surveys. *Journal of Documentation.* 21(3): 169-176; 1965 Sept.

15. Rutgers — The State University. Graduate School of Library Service. Bureau of Information Sciences Research. *Bibliography of Research Relating to the Communication of Scientific and Technical Information.* New Brunswick: Rutgers University Press; 1967. 732 p.

16. Menzel, Herbert. Information needs and uses in science and technology. In: Cuadra, Carlos A., ed. *Annual Review of Information Science and Technology.* Volume 1. New York: John Wiley & Sons; 1966; p. 41-69.

17. Herner, Saul and Mary. Information needs and uses in science and technology. In: Cuadra, Carlos A., ed. *Annual Review of Information Science and Technology.* Volume 2. New York: John Wiley & Sons; 1966; p. 1-34.

18. Paisley, William J. Information needs and uses. In: Cuadra, Carlos A., ed. *Annual Review of Information Science and Technology.* Volume 3. Chicago: Encyclopaedia Britannica; 1968; p. 1-30.

19. Allen, Thomas J. Information needs and uses. In: Cuadra, Carlos A.; Luke, Ann W., eds. *Annual Review of Information Science and Technology.* Volume 4. Chicago: Encyclopaedia Britannica; 1969; p. 3-29.

20. Ladendorf, Janice M. Information flow in science, technology and commerce: a review of the concepts of the sixties. *Special Libraries.* 61(5): 215-222; 1970 May-June. (Reprinted in Sherrod, John; and Hodina, Alfred, eds. *Reader in Science Information.* Washington, D. C.: Microcard Editions; 1973; p. 70-75.)

21. Lipetz, Ben-Ami. Information needs and uses. In: Cuadra, Carlos A.; Luke, Ann W. eds. *Annual Review of Information Science and Technology.* Volume 5. Chicago: Encyclopaedia Britannica; 1970; p. 3-32.

22. Crane, Diana. Information needs and uses. In: Cuadra, Carlos A.; Luke, Ann W. eds. *Annual Review of Information Science and Technology.* Volume 6. Chicago: Encyclopaedia Britannica; 1971; p. 3-39.

23. Hills, Jacqueline. *A Review of the Literature on Primary Communications in Science and Technology.* (Aslib Occasional Publication No. 9) London: Aslib; 1972. 36 p.

24. Lin, Nan; Garvey, William D. Information needs and uses. In: Cuadra, Carlos A.; Luke, Ann W., eds. *Annual Review of Information Science and Tech-*

nology. Volume 7. Washington, DC: American Society for Information Science; 1972; p. 5-37.

25. Meadows, A. J. *Communication in Science*. London: Butterworths; 1974. 248 p.

26. Martyn, John. Information needs and uses. In: Cuadra, Carlos A., ed. *Annual Review of Information Science and Technology*. Volume 9. Washington: American Society for Information Science; 1974; p. 3-23.

27. Crawford, Susan. Information needs and uses. In: Williams, Martha E., ed. *Annual Review of Information Science and Technology*. Volume 13. Chicago: Knowledge Industry Publications; 1978; p. 61-81.

28. Dervin, Brenda; Nilan, Michael. Information needs and uses. In: Williams, Martha E., ed. *Annual Review of Information Science and Technology*. Volume 21. Chicago: Knowledge Industry Publications; 1986; p. 1-33.

29. Kremer, Jeannette Marguerite. *Information Flow Among Engineers in a Design Company*. Urbana, Illinois: University of Illinois; 1980. 158 p. Dissertation.

30. Nkereuwem, Edet Efiong. *An Analysis of Information Use by Scientists and Engineers in the Petroleum Industry in Nigeria*. Ann Arbor, MI: University of Michigan; 1984. 184 p. Dissertation.

31. Patterson, C. J. Dissemination strategies for technical and research information in the building industry. In: *To Build and Take Care of What We Have Built With Limited Resources*: Proceedings of the Ninth CIB Congress held in Stockholm on August 15-19, 1983; Volume 5: *Making Use of Building Research*. p. 269-279.

32. Lacy, William B.; Busch, Lawrence. Informal scientific communication in the agricultural sciences. *Information Processing & Management* 19(4): 193-202; 1983.

33. Shuchman, Hedvah L. *Information Transfer in Engineering*. Glastonbury, Connecticut: The Futures Group; 1981. 265 p.

34. Jones Russel C.; LeBold, William K.; Pernicka, Becky J. Keeping up to date and solving problems in engineering. In: *World Conference on Continuing Engineering Education, Proceedings*: IEEE; 1986; p. 784-791.

35. Raitt D. I. The Information needs of scientists and engineers in aerospace. *The Value of Information as an Integral Part of Aerospace and Defence R&D Programmes*. Cheltenham, UK: AGARD; 1986; p. 3.1-3.5. (AGARD Conference Proceedings No. 385.)

36. Pinelli, Thomas E.; Glassman, Myron; Oliu, Walter E.; and Barclay, Rebecca O. *Technical Communications in Aeronautics: Results of an Exploratory Study*. Washington, DC: National Aeronautics and Space Administration. NASA TM-101534, Parts 1-2. February 1989.

37. Pinelli, Thomas E.; Glassman, Myron; Barclay, Rebecca O.; and Walter E. Oliu. *Technical Communications in Aeronautics: Results of an Exploratory Study—An Analysis of Managers' and Nonmanagers' Responses*. Washington, DC: National Aeronautics and Space Administration. NASA TM-101625. August 1989.

38. Pinelli, Thomas E.; Glassman, Myron; Barclay, Rebecca O. and Oliu, Walter E. *Technical Communications in Aeronautics: Results of an Exploratory Study—An Analysis of Profit Managers' and Nonprofit Managers' Responses.* Washington, DC: National Aeronautics and Space Administration. NASA TM-101626. October 1989.

39. Ladendorf, *op. cit.*

40. Peterson, Ivars. The electronic grapevine: the changing face of scientific communication. Unpublished paper presented at the 1990 American Library Association Annual Conference.

41. Lancaster, F. Wilfrid. *Toward Paperless Information Systems.* New York: Academic Press; 1978. 178 p.

42. Allen. *op. cit.* p. 163.

43. Lavendel, Giuliana. Xerox network: a true believer's view. *Science and Technology Libraries.* 8(2): 31-39; 1987/88 Winter.

44. Garvey, William D. *Communication: the Essence of Science.* Oxford: Pergamon Press; 1979: Chapter 5: p. 126.

The Information Quest
as Resolution of Uncertainty:
Some Approaches to the Problem

Pamela C. Sieving

SUMMARY. Several techniques can be observed among scientists seeking to meet specific information needs or coping with the excess of information presented to them. A taxonomy is applied to these techniques, and some suggestions are made for meeting perceived needs.

The ideal user of scientific and technical information: shall we design one? The positive personality traits one would wish for in a spouse would do nicely for a start: patient; thoughtful; caring; able to see our (the librarians') virtues through the fog of early-morning not-yet-wakefulness or late evening exhaustion; able to communicate needs and wants clearly and without a lot of emotional overtones; willing to work hard to obtain desired results; and appreciative of what we (librarians) do. Add to this other talents which, while useful, are less often associated with marriage: the ability to understand and retain complex organizational details; patience with less-than-perfect systems; an intuitive understanding of how information is organized; and a sense of how best to cope with the inter-

Pamela C. Sieving is Director of Library Services at the John W. Henderson Library of the W.K. Kellogg Eye Center, the Department of Ophthalmology at the University of Michigan. Her previous positions were at Yale University, the University of Illinois—Chicago, and Kirkland and Ellis, a law firm in Chicago. She holds a BA degree from Valparaiso University, MA from the University of Wisconsin—Madison, and MS from Southern Connecticut State College, and has been active in the Machine Assisted Reference Section, Reference and Adult Services Division of the American Library Association. She is currently learning Japanese.

face between the deluge of possible information resources and the limited needs of one particular user (not to mention limited capacities for retaining and controlling that information flood once the gates are open).

Doesn't sound like anyone I've met over the years. Bits and pieces, maybe, bits and pieces (or was that only a song running through my mind).

So who are our real, flesh-and-bloodied users? What are they like? What *do* they like? And how do they approach the perpetual Sisyphean task of identifying, locating, recognizing, assimilating, and storing information? Systematization may help us distinguish some rights and wrongs in how we approach our interaction with them, and our preparation — personal and systemic — for that interaction.

Unlike Caesar considering Gaul, we will divide our user population into four parts for convenient control. The following categories should probably be labelled "enlarged and exaggerated to show detail," and will fit no one user in particular, but many with some accuracy.

First, though, a definition: if we are looking at information-seeking behaviors, what is information? One answer is that of Robert W. Lucky[1] in *Silicone Dreams: Information, Man, and Machine*, discussing Claude Shannon's seminal information theory work: "Information is viewed as being the *resolution of uncertainty*." Shannon also cautions, however, that information may "best be described in terms of *organization*," and proposes a pyramid moving upwards from data through information and knowledge to wisdom.[2] (But since wisdom is beyond the scope of this paper, this is added merely for the reader's edification and meditation.)

THE BIG FILERS

Many, particularly younger or less-mature, information-seekers may be placed in my first category, that of *information filers* or *controllers*. They need a copy of *everything* for their files. Perhaps the researcher approaches the problem of identifying, acquiring, organizing and assimilating information, even if subconsciously, as follows:

They hired me as the expert on Y. I know a great deal about Y, of course, having done a thesis and post-doc exploring a minute aspect of the subject. But if they ever find out I don't know EVERYTHING about it, I'm in deep you-know-what. And I was always impressed when Prof. Weisheit went to her files and pulled out a copy of exactly the reference I needed to solve a problem. If I have everything in *my* files, and somehow find time to learn it all, maybe I'll get by.

I suspect that only those working in something akin to a special library setting may be aware of the extent of this approach; many seekers in this category, lost in the sheer impersonal size and complexity of our major university collections, are hidden from our view except in the occasional interaction concerning interlibrary acquisitions or mediated literature searches. The increasing availability of end-user systems, e-mail interlibrary loan requests, and semi-automated campus delivery systems may shield them even further.

We all know some of what is wrong with this approach. We have fielded a request for "everything on tumor necrosis factor," "everything published on carbon black in the last twenty years" or something similar as the basis for an interaction with a user. Perhaps we have used the strategy of pulling out retrospective bibliographies, *Science Citation Index*, and cost estimates for searching 20 years of backfiles of 13 different databases, out of sheer frustration.

And not only is "retrospective" a problem; current awareness may be worse. Bernier and Yerkey's[1] decade-old estimate of the size of the annual increment of the biomedical knowledge base (200 million new articles per year . . . approximately 5500 per day . . . and that's not even looking at *yesterday's* production) is enough to stun anyone. A recent note in *Science*[4] adds further uncertainties to the already fuzzy set of what comprises "all" of a subject: meetings on sequencing the human genome drew "between 50 and 150 attendees at each session" at the fall, 1990, meeting of the *American Chemical Society*. Yes, chemists!

So here is our first major issue: how to assist the user who thinks "the information situation" will be solved by the acquisition of a copy of everything on a subject, when not only is the subject unmanageably large, but most likely not definable in any very comforting way.

ASK FRANK

A second approach is similar in tone, but quite different in practice. Our information seeker in this category exhibits *no* concern about keeping up with the literature or building a knowledge base beyond the immediate issue. Information is important, but uncertainty is resolved in this mode by passing the buck. If the reader is about to dismiss this article as pure fantasy at this point, let me assure the reader that "I ask Frank" was the opening—and nearly closing—response of a sixth-year graduate student at the University of Michigan when queried as to his information-seeking behaviors.

This student has apparently learned the above lesson *too* well. There is simply too much out there: too much data, too many experts, too many theories, perhaps even too many bibliographic file managers for the PC. Someone else can do the work of making sense of it all. Or, they had better, because at least one of us can't or won't!

Bichteler and Ward[s] identified a major reliance on personal contacts with colleagues when seeking specific information, and similar reliance on "personal networks" in keeping current with the literature among geoscientists. These "Franks" may be local or remote, and there may be a formal or informal understanding of this network between colleagues. Interestingly, these coping behaviors are said to increase when libraries are poor or inaccessible, and to decrease, not with the availability of library collections, but rather as dependence on librarians increases:

> They often just describe a new project to the librarian and then return to their offices, waiting for reports, database searches, etc., to appear.

So, who is Frank? He can be anyone: professor; TA; adviser; colleague; fellow student; lab assistant. Yes, even a librarian. Obviously Frank represents someone trusted as an information resource; a slightly more sympathetic Dr. Weisheit, perhaps. But certainly this approach lacks either self-confidence in managing information acquisition, or, just as likely, a lack of knowledge of the implications of information deficits. What you don't know can hurt, badly; that to which you don't have access can't ever be yours. Grants have recently been denied by federal funding agencies when a faculty could not prove ready-enough access to current journals.'

That this leads to several problems relating to the state of the information arteries is not hard to assume. But is it related to overdoses on too-rich information sources, lack of moderation in some form? Or, merely laziness, lack of incentive to engage in information-seeking exercise. It is unlikely we will have many opportunities to address the Frankly-dependent information user, unless we have become Frank ourselves. And then we must become frank . . . enough!

THE OSTRICH

This user has not even a Frank to rely on. The information problem is ignored, assuming: (a) I already know everything I need to know; or (b) I will run across anything of real importance at a conference. Ostriches may be said to have a great tolerance for uncertainty and ambiguity.

No ostriches in your world? There are in mine. A recent study at a neighboring medical center offered free, 24-hour access to MEDLINE/PaperChase to a group of physicians, and controlled by offering free mediated searching to another group. PaperChase is about as friendly as they come: branching logic offers related subject headings, suggests alternate search strategies, interrupts lengthy displays to propose useful ways to limit too-large search results. The system even stores searches (without the user's request that it do so) for six months, so they can be retrieved later.

Four out of the experimental group of 25 physicians made use of PaperChase. Four. In one year.'

There are frequent articles in the medical literature designed to reach various segments of the clinical research community concerning the availability and usefulness of online literature searches, as there are in most scientific areas. Ease at the computer terminal seems not to have been a factor in this study. But like the old aphorism concerning a horse's thirst, the seeker and the sought-for can be brought together just so far, unless you plan to force-water. Information, in this world, must knock on the laboratory door and announce: "Here I am! Use me!" But most librarians do not like to send their information out on uncharted journeys, especially when seemingly plenty of users act as though the journey should be from user to information, rather than the reverse.

THAT CERTAIN USER

Perhaps, after all, there may be users who have managed to steer a course around the above obstacles. Not yet the ideal; but then which of us as midwives (apologies to male readers) in the information process regard ourselves as quite finished, either? We must all admit to knowing many who do quite well: they have limited their comprehensive-information needs to reasonable quanta, so their current-awareness needs and filing space are manageable; they can resolve uncertainty efficiently and with confidence; their personal network is reliable, both in terms of availability and accuracy.

Like the nominator-librarians in the Sabine" study of how books and journals are used, we may after all best remember those who make a definite impression on us, but not be so very accurate with those impressions. (In this study, users of scientific and technical materials in various libraries were to be interviewed; some of the nominees were later disqualified when they stated that they read only fiction.)

Only 16% of the most recent document usages of those interviewed in this study were for something other than a specific information need; those needs and most others were filled in only seven percent of instances by the use of an entire document. No wonder the authors chose to emphasize a plaint not about the identification or acquisition of information, but about the organization of information already identified and acquired:

> These data are not organized necessarily for easy browsing, and one cannot get at just that information one wants . . . we're overloaded with books and reports, and it's hard to find one's way around and through them.

IN CONCLUSION

What have we learned? Shall we be Frank, or angry, or long-suffering, or pro-active? Shall we build our collections, locally mount databases, or increase access to full-text sources? Bibliographic instruction at the university level [my son says: "BI? What do you buy in the Library?] and higher-order information skills starting in elementary school?

> oh if we knew
> if we knew what we needed if we even knew
> the stars would look to us to guide them[9]

Yes.

REFERENCES

1. Lucky, R.W. *Silicon Dreams: Information, Man, and Machines*. New York: St. Martin's Press, 1989; 40.

2. Lucky, R.W. *Silicon Dreams: Information, Man, and Machines*. New York: St. Martin's Press, 1989; 20.

3. Bernier, C.L. and Yerkey, A.N. *Cogent Communication: Overcoming Information Overload*. Westport: Greenwood Press, 1979; 39.

4. Hamilton, David P. Co-opting the chemists. *Science* 1990; 249: 1249.

5. Bichteler, J. and Ward, D. Information-seeking behavior of geoscientists. *Special Libraries* 1989; :169-78

6. Apted, J., ed. *The University of Michigan Library Newsletter* 1990; 12(35): 3.

7. Wolffing, B.K. Computerized literature searching in the ambulatory setting using PaperChase. *Henry Ford Hospital Medical Journal* 1990; 38: 57-61.

8. Sabine, G.A. and Sabine P.L. How people use books and journals. *Library Quarterly* 1986; 56: 399-408.

9. Merwin, W.S. The Different Stars. In: *The Carrier of Ladders*. New York: Atheneum, 1984.

Meeting the Academic
and Research Information Needs
of Scientists and Engineers
in the University Environment

Harry Llull

SUMMARY. Today's research-intensive universities require science and engineering librarians to address both the academic and research information needs of the various members of the university community. This paper defines some of the differences between the information needs of the traditional academic patron and those of the researcher. It proposes that horizontal organizational structures, emphasis on the team concept, and the electronic library environment come together in a synergetic way to assist librarians in providing informational services in an environment of conflicting priorities.

INTRODUCTION

Academic librarianship has experienced a number of changes over the last twenty years. The implementation of new technology in all facets of our operations and the experimentation with less hierarchical organizational structures for matrix management and broader spans of control are just two comprehensive developments that have changed and shaped our internal operations and day to day jobs. However more importantly, the information needs of our faculty and researchers have changed and grown. The primary motivation behind this latter change is the tremendous growth in research

---

Harry Llull is Director of the Centennial Science and Engineering Library at the University of New Mexico in Albuquerque, NM. He received his BS from Auburn University, Alabama and his AMLS from the University of Michigan. The Centennial Science and Engineering Library was opened on February 1, 1988, and is part of the General Library System at UNM.

83

and development money from the federal government to the universities. Many universities have become science research centers to a greater or lesser degree.[1] Academic librarians must address these changes in terms of the information needs of researchers on contracts and grants while also continuing to supply and address the information needs of the students and instructional role of the faculty.

TRADITIONAL AND RESEARCH-INTENSIVE ACADEMIC ENVIRONMENTS

First I would like to define what I refer to as the traditional academic environment. That environment is one which emphasizes the learning or discovery mode motivated by the individual's commitment to expand the human knowledge base. The environment is curriculum based. Library collections are expected to be comprehensive in the areas of instruction so that ideas and concepts may be researched step by step. Comprehensive collections also allow for the discovery of knowledge by serendipity. The library takes on the role of an institution within the broader concept of a university with the library building and the collections it contains being the physical embodiment of that role. In some cases, this view has also generated the less flattering view of librarians as the caretakers of those collections. Our patron in this traditional mode is the generic student. Students may fall into various categories: undergraduates, graduate, faculty, and the public. All come to the library to explore the wealth of knowledge. Academic librarians have seen their role as one to facilitate that exploration. Because the academic environment emphasizes the process of that exploration as much as the results, time pressures may be more flexible. Academic librarians tend to emphasize group instruction for classes to teach the basic library skills that would be relevant for a variety of courses.

The role universities have taken on as major research and development centers for the government and industry has added to the academic librarian's information role. "The 'new' university possesses institutional features which are appropriate to its broadened functions, i.e., integration of science and technology. It means primarily, that the traditional system of disciplinary departments (both basic and applied) is supplemented by non-disciplinary, technology oriented units and research programmes. The function of the latter is to undertake R&D normally not suitable for disciplinary departments and to develop appropriate mechanisms for closer interactions with various clients and external R&D environments (in industry, government, etc.).''[2] The amount of monies involved is significant. In the past, the top 100 funded universities have shown ranges of funding from $465,453,000 to Johns Hopkins University to $27,377,000 to Auburn University, Alabama.[3] In addition, the Carnegie classification of a university includes the amount received annually in federal support for research and development as a criteria for deciding into which category a university will fall. In 1987, $33.5 million or more helped define research universities I while $12.5-$33.5 million defined research universities II. Other categories such as doctorate-granting universities did not have a monetary criteria for federal support for research and development although many such universities also receive funding.[4] The new university environment presented here is one with a "much more active partnership with industry than we have seen in the past. Universities must reach out. Pragmatism has been America's ideology. We must learn again to wed effectively the dreamers and the knowledge makers with knowledge brokers and practitioners. The stakes are America's future."[5]

These closer ties with both industry and government have raised some concerns in terms of the traditional role of a university. "At many universities the level of research has grown so that it has created explosive pressures on space, faculty, money and time which can't help but affect costs . . . University research has helped expand the state of our knowledge. In an economy driven by technology, it continues to be vital. But why must we put support for teaching and research in the same budget?" At the center of the controversy is the overhead costs the federal government allows to universities. The percentage of overhead a university charges on a. federal grant or contract can vary widely from university to university. On the one hand, as overhead charges go up there is less of a pool of money for additional grants.[7] On the other hand, many universities feel they are under-reimbursed and must use their own money to finance and manage government research.[8] Whether in universities supported by state monies or in private institutions with high tuition costs to students, academic librarians have the challenge of meeting the information needs of all groups within the university within the restraints of budgets, time, and human resources.

Next we need to define the new research-intensive academic environment and how the information needs differ not only from the perception of the researcher but our perceptions of ourselves as information providers. One very important fact is that the research undertaken at a university may not be curriculum based. So the first question is can we afford comprehensive collections both in terms of buying and storing materials needed by researchers on grants and contracts? What are the implications on collections when a short-term research project is awarded in an area that the university has not emphasized in the past and the library has not built up a research collection? Because of the overhead issue, researchers do expect to be able to turn to the library for their information needs. Compounding the issue is the fact that it is the science and technology literature that is primarily needed for this research and is the vary body of literature that at present has had the greatest impact on library collection development budgets because of the annual increases in prices. These researchers also often need specialized materials and technical reports that are not always part of the collection development policy of university libraries. All these factors call upon the library to provide online services that will help the researcher to identify comprehensively what has been published. The online services must be backed up by a timely document delivery service for those materials not in the collection. Most importantly, the librarian needs to be viewed as a member of the research team and available for addressing intensive information needs of the research group. The librarian becomes the focal point of the research team for its information needs as opposed to the university library building and the collections it contains. The primary clientele of the research team and the librarian as a member of that team is the government or industrial agency that assigned the grant or contract. Time becomes paramount and library instructional services become much less important. The primary goal is to complete the research project on time.

AN ORGANIZATIONAL RESPONSE

Effectively managing the human resources one has been allocated is a key factor in meeting both the traditional academic information needs and the information needs of researchers on contracts and grants. There are three concepts and approaches that when implemented together create a productive synergetic environment in terms of the delivery of information services. These concepts are: nonhierarchical or horizontal organizational structure; teamwork, both within the library and with the researchers and academic departments; and the electronic library.

"A healthy organization is one that adapts to the needs of the employees, as well as to the pressures of its environment. Such an organization is characterized as 'organic' and it considers wide participation in decision making, and emphasis on mutual dependence of the employees and cooperation is very important . . . An open organization is defined as one that accepts input of energy from its environment and adapts itself to better suit that environment in order to increase future inputs of energy."[9] In referring to our organizational environment, we need to think not only of our internal library organization but also view our clients as part of our organization who need to have input into what services are provided. Library organizations need to be "evolutionary, nonheirarchical, entrepreneurial, and horizontal . . . This growth sequence will require librarians to become more involved in the process of scholarly communication . . . Networks are people talking to each other, sharing information, ideas, and resources . . . We will restructure our businesses into smaller and smaller, more entrepreneurial, more participatory units."[10]

With an horizontal organizational structure, emphasizing participation and input by those involved and affected by organizational priorities and decisions, teamwork and the team concept becomes a natural progression. Teamwork both within the library organization and with the academic departments and research groups is critical. "Librarians and researchers must form partnerships in order to facilitate the research process . . . Each social group has its own culture, including value system, which affects the patterns of creating, recording, producing, disseminating, organizing, diffusing and preserving information and knowledge.''[11] It is critical that librarians work closely with the research team in order to gain a better understanding of the creative process for a particular discipline. "When solving problems and building theories, researchers seldom operate in a carefully calculated, linear, sequential pattern.''[12] Librarians need to assert themselves and anticipate the needs of the researcher and become a visible member of the research team. "The potential benefits of the information specialist as an active team member in the research process have only begun to be explored. There is a need for personalized, sympathetic, and knowledgeable information counseling in all disciplines and in all organizations. The future beckons brightly for anyone willing to assume this role.''[13]

The electronic library, primarily through electronic mail over campus networks, can provide an avenue of input to organizational decisions from our clientele and add to the inclusion of the librarian as part of the research team outside of the library. The following are four attributes of the electronic library: "management of resources with a computer; the ability to link the information provider with the information seeker via electronic channels; the ability for staff to intervene in the electronic transaction when requested by the information seeker; and the ability to store, organize and transmit information to the information seeker via electronic channels.''[14] Electronic communication not only is more efficient but it helps break down the organizational barriers of being in different buildings and members of different departments within the university. "Library as a place will give way to 'library' as a transparent knowledge network providing specialized librarians and energizing information technology.''[15] The librarian's workstation will "facilitate electronic information imaging, publishing, telecommunications, and information delivery in addition to networked collection management and reference services.''[16] Because of the amount of information available and the ease of access to it electronically, the role of the librarian in terms of evaluating the information becomes more important. "The gatekeeper is an individual who controls the flow of communication to other people. Because of their authority and skills, gatekeepers are in a position to influence others through the amount of correct or incorrect information that they allow through . . . The librarian can also be viewed as an information transfer agent and should serve as an advocate to access the information."[17]

CONCLUSION

The challenge of meeting both academic informational needs and research-intensive informational needs within an academic environment has made academic librarianship even more exciting. Many academic librarians find that access to campus-wide networks, Bitnet, and Internet provides them with the necessary electronic access to researchers and information on their own campuses and throughout the world. Electronic networks have provided academic librarians with electronic colleagues with whom they can share information, ideas, and concerns. While there is still a need for librarians to be able to handle information queries that one will encounter at the traditional academic library reference desk, it is also important that librarians be assigned to specific departments in order to develop the understanding and skills to address a specific discipline's specialized reference needs, online searching requests, and collection development or material access requirements. As librarians take on more of these specialized needs, support staff need to be of a higher level of training and skills in order to assist in the delivery of services to the undergraduate including providing instructional services. The organization will require the flexibility to respond to constantly changing workloads and priorities. This again is enhanced with an horizonal organizational structure and emphasis on teamwork. This type of organizational structure is one that will continue to go through an evolutionary process. That fact is the very foundation of the horizontal structure and team concept. The electronic networking increases the level of access from staff and clientele which creates the need for evolutionary changes within the organization, providing a stimulating and challenging environment within which to work.

REFERENCES

1. McDonald, Kim. NSF to launch 10 to 20 university science-research centers. *Chronicle of Higher Education*. 34 (5): A7; 1987 Sept. 30.

2. Stankiewicz, Rikard. *Academics and entrepreneurs; developing university-industry relations*. London:Frances Pinters; 1986. p.114.

3. U.S. funds for college and universities. Chronicle of Higher Education. 34 (15): A22; 1987 Dec. 9.

4. How classifications were determined: text of the category definitions. *Chronicle of Higher Education*. 33 (43): 22; 1987 July 8.

5. Newell, Barbara W. The Future of the state universities: continuing education and research in an era of science-based industries. *In*: Koepplin, Leslie W.; Wilson, David A., ed. *The future of state universities; issues in teaching, research, and public service*. New Brunswick, New Jersey: Rutgers University Press; 1985. pp.102-103.

6. Honigs, David E. Why college costs go up. *Newsweek*. 111 (23): 8; 1988 June 6.

7. Cordes, Collen. Some Stanford researchers fear high overhead costs will curb grants. *Chronicle of Higher Education*. 34 (33): A25; 1988 April 27.

8. McDonald, Kim. U.S. under-reimbursement forces universities to pay millions to manage research, report says. *Chronicle of Higher Education*. 34 (11): A23; 1987 Nov. 11.

9. Dowlin, Kenneth E. *The Electronic library; the promise and the process*. N.Y.: Neal-Schuman Publishers, Inc.; 1984. p.44.

10. Gapen, D. Kaye. Myths and realities: university libraries. *College & Research Libraries*. 45 (5): 350-361; 1984 Sept.

11. Grover, Robert; Hale, Martha L. The role of the librarian in faculty research. *College & Research Libraries*. 49 (1): 9-23; 1988 Jan.

12. Ibid.

13. Neway, Julie M. *Information specialist as team player in the research process*. Westport, Conn.:Greenwood Press; 1985. p.162.

14. Dowlin, p.33.

15. Murr, Lawrence E.; Williams, James B. The roles of the future library. *Library Hi Tech*. 5 (3): p.7; 1987 Fall.

16. Ibid.

17. Dowlin, pp. 36-37.

SPECIAL PAPERS

Physical Structure and Administration of Science and Technology Libraries: An Historical Survey

Elizabeth P. Roberts
Elaine Brekke
Kimberly Douglas

SUMMARY. This article discusses the results of a survey tracing the evolution of the physical and administrative structures of science and technology libraries in 75 ARL institutions in the United States and Canada from the 1940's through the 1980's. The physical structure of science and technology libraries is compared on the basis of private vs. public universities, age, size, and geographic location.

INTRODUCTION

Historically, science and technology (sci/tech) libraries have been mavericks. University administrators and directors of libraries have been challenged with how to manage them and for years many were left alone to the de facto administration of academic depart-

Elizabeth P. Roberts is Head, Owen Science and Engineering Library, Washington State University.

Elaine Brekke is Reference Librarian, Owen Science and Engineering Library, Washington State University.

Kimberly Douglas is Head, Reader Services, California Institute of Technology.

ments. In the early days, not only did departmental/branch libraries flourish, they multiplied rapidly. Physical configurations have run the gamut from Harvard University where, in 1982, there were 90 different science and engineering library units, many operated by teaching departments, to Johns Hopkins University where centralization has occurred with only a main library and one branch.[1] Most universities are somewhere between these two extremes.

At the end of the heady days of the 1960's, many university libraries discovered that they could no longer afford to sustain the 5 or 6 copies of *Chemical Abstracts* located in various departmental libraries on campus; they could no longer afford to pay for professional staff at numerous locations. It became evident that it was more cost effective to deliver photocopies of articles and books to departmental offices rather than to maintain a small library in close proximity to the department in question. Faculty and students in interdisciplinary subjects and librarians hailed the arrival of the centralized sci/tech library with alacrity, while those who had benefited from a library down the hall from their offices did not share in the enthusiasm of their colleagues. In the long run, economics ruled the day and we witnessed the birth and multiplication of the centralized sci/tech library.

THE SURVEY

In order to check the validity of the above statements, the Committee on the Comparison of Science and Technology Libraries of the Science and Technology Section (STS) of the Association of College and Research Libraries undertook a survey. Previously, members of this committee had developed a statistical survey to gather data to be used as a baseline for comparing various service aspects of sci/tech libraries.[2] Information gathered from this survey on various aspects of the operation of these libraries proved to be the first descriptive data available for comparing the activities taking place. Another study[3] summarized data on the existing physical and administrative structures of sci/tech libraries. The data were collected by the same STS committee.

This survey was sent to 118 libraries, with ninety-seven libraries responding (82%). For purposes of comparison, all eight non-ARL libraries were eliminated as were seven ARL libraries with no history of having a separate sci/tech collection over the period studied. The survey is an effort to discover the trends in both the physical arrangement and the administrative structure of academic sci/tech libraries from the 1940's through the 1980's.

PHYSICAL STRUCTURE OF SCI/TECH LIBRARIES

This part of the questionnaire was designed to determine if a university housed the sci/tech collection in a principal sci/tech library, in multiple branch/departmental libraries, or maintained a collection completely integrated into the main library's collection. Participants were asked to respond with data for the years from the 1940's through the 1980's. (See Figure 1.) Results of the survey are summarized in Table 1.

The results of the survey show that between 1940 and 1980 there is a steady decline in the number of institutions with multiple branch/departmental libraries and a corresponding increase in the number of centralized sci/tech libraries. Libraries with science and technology collections integrated into the main collection remained at 8 or below during the years studied. If reporting had been complete for 1940, one can speculate that the figures might have been higher for integrated collections, and for multiple branch/departmental libraries.

PHYSICAL STRUCTURE PRIVATE VS. PUBLIC UNIVERSITIES

Multiple branch/departmental libraries have been slower to disappear in privately funded universities than they have in those which are publicly funded. In publicly funded universities, the number of centralized sci/tech libraries has steadily increased from the 1940's through the 1980's, with a correlating decline in multiple branch/departmental libraries. (See Figures 2A & 2B.)

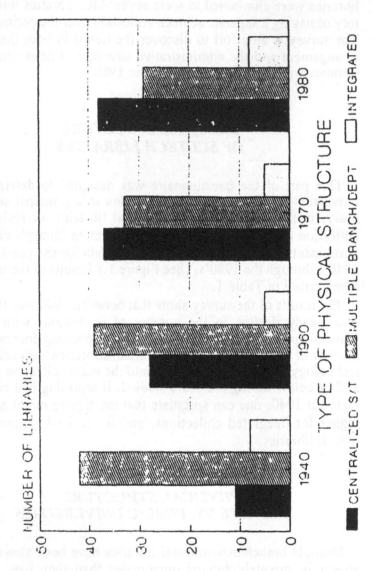

FIGURE 1

PHYSICAL STRUCTURE S/T LIBRARIES

PHYSICAL STRUCTURE BY AGE OF UNIVERSITY

Universities were divided into 3 groups — Group I, the oldest 25 universities (founded 1850-1872), Group II, the middle 25 (1850-1872) and Group III, the youngest 25 (1873-1957). The decentralized configuration of many branch/departmental libraries shows a steady decline in both Group I and II, while remaining about the

TABLE 1. Physical Structure (See Figure 1.)

	CENTRALIZED SCI/TECH	MULTIPLE BRANCH/ DEPT.	INTEGRATED INTO MAIN	NO REPORT
1940	11	42	8	14
1960	28	39	5	3
1970	37	33	5	0
1980	38	29	8	0

NOTE: Some libraries were unable to supply the information for the early years.

FIGURE 2A

PHYSICAL STRUCTURE PRIVATE UNIVERSITIES S/T LIBRARIES

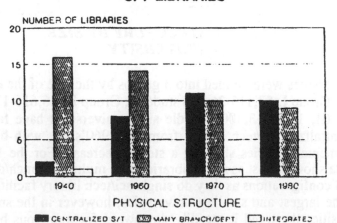

NUMBER OF LIBRARIES

PHYSICAL STRUCTURE

■ CENTRALIZED S/T ▨ MANY BRANCH/DEPT ☐ INTEGRATED

FIGURE 2B

PHYSICAL STRUCTURE PUBLIC UNIVERSITIES
S/T LIBRARIES

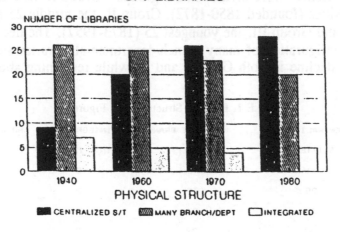

same in the 25 youngest universities (Group III). From the 1970's on, centralized sci/tech libraries are the predominant type of library in the Group II institutions and by far the configuration of choice in the younger institutions from the 1960's on. (Source of age of university: *HEP Higher Education Directory.* Washington, Higher Education Publications, 1989.) (See Figure 3.)

PHYSICAL STRUCTURE BY SIZE OF UNIVERSITY

Universities were divided into 3 groups by the size of the enrollment — Group I, 53,115-26,616 students; Group II, 26,616-17,265; Group III, 17,166-3,100. Middle sized universities have favored the centralized sci/tech library from the 1960's on, with branch/departmental libraries showing a steady decrease. For the 1980's the data show almost as many libraries with multiple branch/departmental configurations as they do single sci/tech library facilities for both the largest and smallest universities, however in the smallest universities the trend is steadily downward for numerous branch/

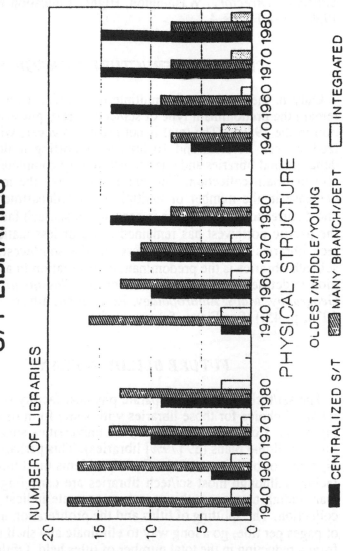

FIGURE 3

PHYSICAL STRUCTURE BY AGE OF UNIV. S/T LIBRARIES

departmental libraries. (Source of size of university: *HEP Higher Education Directory*. Washington, Higher Education Publications, 1989.) (See Figure 4.)

PHYSICAL STRUCTURE BY REGION

Only in the northeast do multiple branch/departmental libraries remain the predominant type of sci/tech library physical configuration in the 1980's. The total is not great, however, with 8 centralized sci/tech libraries and 10 libraries reporting multiple branch/departmental libraries and one sci/tech library completely integrated into the main collection. The south has shown the most dramatic decrease in the number of multiple branch/departmental libraries and a corresponding increase in centralized sci/tech facilities, while the pacific northwest has remained more or less static with 2 sci/tech libraries throughout the years. In the southwest, centralized sci/tech libraries is the predominant configuration in all years. (Region is defined by, Library of Congress, *Classification, Class G, Geography, Maps, Anthropology, Recreation*, 4th ed. Washington, 1976, p. 45.) (See Figure 5.)

FUTURE BUILDING PLANS

The sci/tech library as a separate physical facility is the configuration of choice for those libraries with some form of future building plans. However, the majority of universities surveyed has no future building plans (59 [79%] libraries). This perhaps reflects the high cost of building, and the fiscal problems most universities are facing. Although most sci/tech libraries are canceling titles rather than adding new ones, this is not an accurate indicator of a static collection. The splitting of titles and the proliferation in the number of pages per title, go a long way to eliminate any shelf space gained from a reduction in the total number of titles held. Of those libraries which do have future building plans, six (8%) will build a new sci/tech library facility, five (7%) will increase the size of the existing sci/tech library, three (4%) plan branch expansion and two (3%) libraries plan to move into a remodeled centralized facility but keep

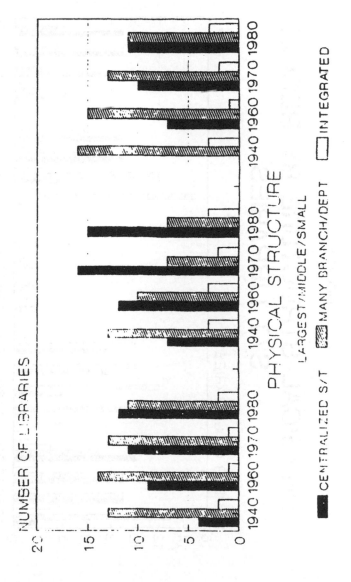

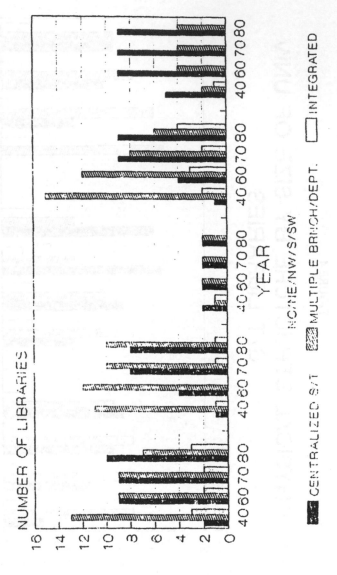

FIGURE 5

PHYSICAL STRUCTURE BY REGION
S/T LIBRARIES

the sci/tech library as a separate entity within the main library. (See Figure 6.)

ADMINISTRATIVE STRUCTURE
SCI/TECH LIBRARIES

Results of the survey are seen in Table 2. They show a steady increase in the administrative configuration where there is one head for the sci/tech library, (from 6, 1940 to 41 in 1980), and a corresponding decrease (from 34 in 1960 to 20 in 1980) in the number with many branch/departmental library heads. The configuration with two or more heads of sci/tech libraries reached a peak in 1970 while no administrative head of this type of library peaked in 1940 (NOTE: Only 75% of the libraries were able to report the administrative structure 1940), and then dropped to 4 in 1970 and 6 in 1980. The administrative structure of sci/tech libraries, not surprisingly, follows a similar pattern to that of centralized sci/tech library facilities. The number of institutions with one sci/tech library administrator shows an increase from the 40's to the 80's, while those having multiple heads of departmental/branch libraries as the dominant administrative configuration of administration of sci/tech libraries have diminished since 1960. Universities indicating no sci/tech library head have decreased from 22 in 1940's to 6 in the 1980's. (See Figure 7.)

It is interesting to note that in the 1980's more libraries have centralized the management of sci/tech libraries than have centralized the physical resources. The former does not generally require faculty support nor is it as capital intensive.

THE FUTURE

Clearly the trend over the last forty years has been to centralize the physical resources for science and engineering disciplines. The identifiable holdouts from this trend are the oldest universities in the country, mostly located in the northeast. Factors probably contributing to this trend have been budget constraints, especially the high cost of duplicate subscriptions for journals, indexes, and for staff; increased need for security for costly science collections; space oc-

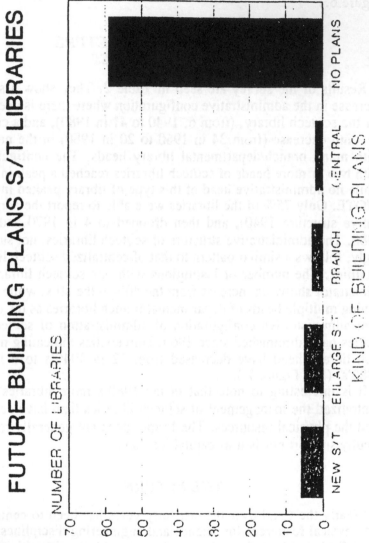

FIGURE 6

FUTURE BUILDING PLANS S/T LIBRARIES

TABLE 2

	One Head of S/E	2 or + Heads	Many Branch/Dept Heads	None	Not Reporting
1940	6	2	26	22	19
1960	19	6	34	10	6
1970	30	11	28	4	2
1980	41	8	20	6	0

NOTE: Some libraries were unable to supply the information for the early years.

cupied by departmental libraries needed by growing departments; and the proliferation of interdisciplinary studies.

Whether this move to centralize collections into a separate sci/tech library facility will continue is one of interest. Automation, while initially an incentive for centralization in some institutions, is conducive to decentralized data input and use. Library material can now be located in numerous remote locations, with records input into a shared centralized database which can be accessed from home, office or branch/departmental library. The increased availability of journals in electronic format could eliminate much of the need for duplicate subscriptions as can the use of fax machines when they are located at the site of the sci/tech collection(s).

Faculty pressure for convenience may also contribute to decentralization and no prudent librarian can dismiss faculty opinion. Ellis Mount states this well:

> Of all the user groups at a college or university, faculty members are the most cohesive, and are capable of making a major impression on a library's style of operation. If faculty members are pleased with the library, they may not be heard from. If they are not pleased, their influence is often such that they can remedy situations to their liking.[1]

Much of the scientific community favors the small library conveniently located near the academic departments.[4] The following quote by a physicist illustrates this point:

FIGURE 7

ADMINISTRATIVE STRUCTURE
S/T LIBRARIES

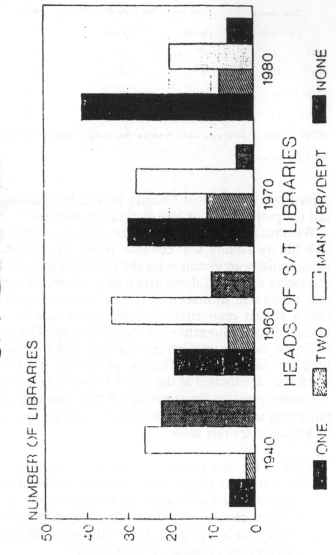

The advantages to be derived from departmental libraries are so obvious I am rather surprised your institution is even considering a change from that system . . . We do not know who is responsible for the idea of a central science library, but presumably it is someone who does not really understand how a science library is used . . . For convenience, for usefulness, for making a contribution to the work of the department there is no substitute for the departmental library, provided a competent librarian is available to service it.[5]

While most library personnel would not like to see a movement back to departmental/branch collections, the future trend is unpredictable and will be influenced by many opposing factors, some of which have been outlined above. It would be useful to follow-up on those institutions that chose centralization and identify the factors influencing the decision to centralize. Such information might provide a basis for predicting the future.

NOTES

1. Mount, Ellis. *University Science and Engineering Libraries*. 2d ed. Westport, Conn., Greenwood Press, 1985. (Contributions in Librarianship and Information Science, Number 49.)

2. Emerson Hilker, "Statistical Data for Stand-Alone Science/Engineering Libraries in the United States and Canada 1984/1985," *Science and Technology Libraries* 8:89-127 (Fall 1987).

3. Elaine Brekke, Kimberly Douglas, Elizabeth Roberts, "Survey of Science and Technology University Library Facilities and Administration," (Submitted for publication, 1990.)

4. Marianne Cooper, "Organizational Patterns of Academic Science Libraries," *College and Research Libraries* 29:357-363 (September 1968).

5. D.A. Wells, "Individual Departmental Libraries vs. Consolidated Science Libraries," *Physics Today* 14:40-41 (May 1961).

Academic Science
and Technology Libraries:
Facilities and Administration

Elaine Brekke
Kimberly Douglas
Elizabeth Roberts

SUMMARY. This article discusses data on the physical and administrative structures of science and technology libraries which were collected by the Committee on the Comparison of Science and Technology Libraries, a standing committee of the Science and Technology section of the Association of College and Research Libraries. The collection of descriptive data was undertaken to determine what physical and administrative structures were prevalent among science/technology libraries. This information will serve as a basis for grouping and comparing the statistical data that the Committee has compiled from previous surveys.

I. INTRODUCTION

The Committee on the Comparison of Science and Technology Libraries,[1] a standing committee of the Science and Technology Section of the Association of College and Research Libraries, was

Elaine Brekke is Reference Librarian at the Owen Science and Engineering Library at Washington State University, Pullman, WA 99164-3200. She received a BA from Douglass College, New Brunswick, NJ and an MLS degree from the University of Wisconsin-Madison.

Kimberly Douglas is Head of Reader Services at the Robert A. Millikan Memorial Library, California Institute of Technology, Pasadena, CA 91125. She received an MA from Freie Universita (Berlin) and an MS degree from Long Island University.

Elizabeth Roberts is Head of the Owen Science and Engineering Library at Washington State University, Pullman, WA 99164-3200. She received a BA from the University of Missouri and an MLS degree from Emory University.

charged to collect, analyze, and distribute data on the organization and operations of academic science/technology libraries. Medical libraries were not included in the surveys because similar statistical data are collected by the Medical Library Association. During the past several years, the Committee surveyed member institutions of the Association of Research Libraries (ARL). The institutions surveyed were limited to those with sci/tech collections housed in a "stand-alone" or "free-standing" facility or with significant sci/tech collections that were housed within a main library. Data were gathered on physical characteristics, clientele, collections, expenditures, personnel, circulation, interlibrary loan, reference, library instruction, database searching, fees, fines, and professional salaries. Data from the first statistical survey were summarized by Emerson Hilker[2] in 1987.

Although the original survey questionnaire was used to collect data which described the physical facility, institutions were given only three broad definitions to use to identify the type of physical facility: (1) a stand-alone sci/tech library, housed in a separate building; (2) a main building which also housed other units; and (3) libraries sharing a building with academic departments. These designations were not specific enough to provide complete information about the physical configurations of these libraries although they provided minimal information about the physical structure. However, there were no options that described the administrative structure. Therefore, in 1986, the Committee designed a new questionnaire that would be used to collect more detailed information about the physical structures and the administrative configurations of sci/tech libraries, and to identify those libraries whose sci/tech collections were not separate but were integrated into the main collection.

II. RESULTS AND DISCUSSION

A number of questions were developed that would gather information describing the physical configurations of sci/tech libraries. Options were provided which described separate sci/tech library facilities (stand-alone), stand-alone sci/tech libraries with and without branch libraries, sci/tech collections housed in a separate location within a main library, and completely integrated collections. Options describing administrative configurations ranged from a single administrator to none at all. A copy of the complete survey questionnaire is included in Appendix I.

One hundred seven U. S. and Canadian member libraries of the Association of Research Libraries (ARL) were sent copies of the questionnaire. Ninety-one libraries (85%) returned completed questionnaires. Data supplied cover 1984-1989. Although questionnaires may be carefully designed, there are cases where the actual situation does not neatly fit within any of the survey categories provided. In order to deal with this situation, each respondent was contacted to verify data placement within a category.

The data were divided into three groups based on *Rank Order Table I: Volumes in Library* of the ARL statistics.[3] This was done to determine if there was a relationship between the size of the library and the physical or administrative structures of those libraries. Group I libraries (25) includes those with collections of 3 million volumes or more; Group II (29) includes those with collections of 2 million volumes or more, and III (37) includes those with total volumes of one million volumes or more.

A. Physical Structure

Table 1 is a tabulation of the data describing the physical structure of the sci/tech libraries responding to the questionnaire. Data based on Questions IA1-IA5 of the questionnaire describe stand-alone sci/tech libraries that are housed separately (42). Eleven (44% of Group I), 16 (55% of Group II, and 15 (41% of Group III) libraries reported some type of stand-alone facility.

Another method of centralizing a library's sci/tech functions is to house sci/tech materials as a separate collection within a main library building (IA6). No libraries in Group I reported this configuration. However, this type of arrangement exists in 1 library from Group II, and in 8 libraries from Group III. Disbursed configurations for sci/tech functions include multiple branch libraries without a main sci/tech library (IB1). Responses in this category included 14 (56% of Group I), 5 (17% of Group II), and 6 (16% of Group III). Five libraries (17% of Group II), 8 libraries (22% of Group III), and no Group I libraries reported sci/tech collections integrated

TABLE 1
PHYSICAL STRUCTURE

QUESTION*	GROUP I > 3 million volumes (N=25)	GROUP II > 2 million volumes (N=29)	GROUP III > 1 million volumes (N=37)
IA1-1 principal sci/tech library	1	5	5
IA2-Sci/tech library library w/independent branches	0	0	2
IA3-Sci/tech library w/dependent branches	8	7	6
IA4-Mixture of IA2 and IA3	1	1	1
IA5-More than 1 sci/tech library	1	3	1
IA6-Separate collection within main library	0	1	8
IB1-Dispersed, multiple branch libraries	14	5	6
IB2-Integrated collection	0	6	8
IC-Other	0	1	0

* Please refer to Appendix I for complete questions.

into the main collection (IB2). One library from Group II reported a configuration which placed it in the "Other" category (IC). This library has an integrated collection with the exception of one branch library which has one specific subject specialization.

One library from Group I, 5 libraries from Group II, and 5 libraries from Group III reported one stand-alone sci/tech facility (IA1). Only 2 libraries (Group III) reported a 1 main sci/tech library with independent branch libraries (IA2). Eight (32% Group I), 7 (24% Group II), and 6 (16% Group III) libraries reported one principal sci/tech library with one or more branch libraries reporting to the main sci/tech library (IA3—dependent). One library from each of the three groups reported a configuration (IA4) which was a mixture of the descriptions from questions IA2 and IA3. A total of 5 institutions reported more than one stand-alone sci/tech facility

(IA5 — 1 or 4% of Group I, 3 or 10% of Group II, and 1 or 3% of Group III).

The two most common configurations for stand-alone facilities for all groups were a separate sci/tech library with no branch libraries (IA1) and a separate sci/tech library with dependent branches (IA3). However, a dispersed arrangement (IB1) is the most common configuration for Group I institutions (56% Group I). The most common subject specialization of the branch libraries, listed in descending order of frequency mentioned, were Mathematics, Chemistry, Engineering, Physics, Geology, and Biological Sciences.

B. Administrative Structure

The most prevalent administrative configuration for all three groups of sci/tech libraries was one administrative head (II1 — 43 or 47% of the 91 libraries responding). The data are shown in Table 2. Eleven (44% of Group I), 14 (48% of Group II), and 18 (49% of

TABLE 2
ADMINISTRATIVE STRUCTURE

QUESTION*	GROUP I > 3 million volumes (N=25)	GROUP II > 2 million volumes (N=29)	GROUP III > 1 million volumes (N=37)
II1-One head	11	14	18
II2-More than head	0	3	3
II3a-Many heads report to central library	2	3	5
II3b-Head reports to coordinator	4	2	0
II3c-Head reports to academic dept.	0	0	0
II3d-Other	7	2	2
II4-No designated administrator	1	2	2
II5-Other	0	3	4

*Please refer to Appendix I for complete questions.

Group III) reported this type of configuration. It is interesting to compare this type of administrative configuration with the data on the physical facility. Although there is a modest increase in the proportion of libraries reporting this configuration from Group I to Group III (from 44% to 49%), this increase does not indicate that the administrative structure is dependent on either the physical structure or the size of the institution.

The data indicate that some institutions centralize administration of sci/tech collections and reference services even though the collections of those institutions may be dispersed, that is, not housed in a stand-alone facility or as a separate collection within a main library. The larger institutions historically have favored a dispersed physical arrangement based on a tradition of strong departmental control and a financial climate that could support multiple library units.[4] Thirteen Group I libraries (52%), 7 Group II (24%), and 7 Group III (19%) report multiple branch library heads who may report to a sci/tech coordinator, unit head, dean, or director located in the central library (II3a – II3d). Although some libraries report to an academic department or liaison, direct or indirect reporting channels to a sci/tech coordinator or administrator in the central library are also in place. These libraries responded to question II3d (Other) rather than II3c (Library heads reporting to individual academic departments). This would indicate that the large institutions have recognized the need to centralize administrative management of sci/tech services, thereby ensuring the uniform application of policies.[5] Furthermore, administrative changes can easily be made by library administrators without major budgetary changes. Although changes in physical facilities may be desirable, administrators must find funds outside of the library's budget before proceeding with new construction or remodeling projects.

The last category of libraries has no designated administrator for sci/tech services (II4). A small number of libraries (or 4% of Group I, 3 or 7% of Group II, 5 or 13% of Group III) reported this arrangement.

There is no clear pattern indicated by libraries who marked question II5 (Other) on the questionnaire. In several cases, the administration of sci/tech functions with the ultimate responsibility of a Director of Public Services or Administrative Services or an Assistant University Librarian for Branch Libraries, but no conclusions could be made based on the data. Placing the management of sci/tech functions in these libraries under one administrator does represent a move towards administrative centralization, but it does not appear to be driven by a concern to centralize management of only sci/tech functions.

III. CONCLUSION

The data collected during this study describe the physical and administrative configurations of sci/tech libraries. The data show that most libraries with holdings of 3 million or more volumes (56% of Group I) have dispersed physical structures. However, some Group I institutions (13 or 52%) with this configuration have achieved some administrative centralization by having those dispersed libraries report to a sci/tech coordinator located in either the main library or in one of the branches. Slightly more than half of the medium-size libraries (55% of Group II) have some type of stand-alone sci/tech library(ies) and a correspondingly high percentage of administrative centralization (58%). The smaller institutions (41% of Group III) also show a high percentage of stand-alone facilities and administrative centralization (56%). Although the data do not show a clear trend toward the establishment of stand-alone facilities regardless of the size of the library, they do seem to indicate a move toward administrative centralization independent of the size or physical configuration of the library.

The data gathered by means of this questionnaire help to describe the most common physical and administrative configurations of sci/tech libraries. This information is considered to be a beginning in the effort to gather descriptive data on sci/tech libraries. Currently, the Committee has completed an historical survey[6] which provides clearer information about trends in the types of physical facilities and administrative configurations of sci/tech libraries. The results of both of these surveys will be used to group institutions and thereby allow a more meaningful comparison of the statistical data previously collected by the Committee.

NOTES

1. The Committee was renamed during the American Library Association Midwinter Meeting in January 1989. The former name was Committee on Comparison of Science and Engineering Libraries.
2. Hilker, Emerson. Statistical Data for Stand-Alone Science/Engineering Libraries in the United States and Canada 1984/1985. *Science & Technology Libraries*. 8:89-127; 1987 Fall.
3. Daval, Nicola; Feather, Celeste. *ARL Statistics, 1987-1988*. Washington: Association of Research Libraries; 1989, p. 42.
4. Cooper, Marianne. Organizational Patterns of Academic Science Libraries. *College and Research Libraries*. 29:357-363; 1968 September.
5. Ibid., p. 359.
6. Roberts, Elizabeth; Brekke, Elaine; Douglas, Kimberly. Physical Structure and Administration of Science and Technology Libraries: An Historical Survey. *Science & Technology Libraries* 11(3) Spring, 1991. p. (107), in this issue.

APPENDIX I

ACRL - SCIENCE AND TECHNOLOGY SECTION

UNIVERSITY SCIENCE & ENGINEERING LIBRARIES: DEFINITION SURVEY

INSTRUCTIONS: Please check the blank that most closely describes the science and engineering library configuration on your campus. Do not include medical libraries; Statistics for medical libraries are gathered by other professional organizations. Answer for a single campus only. e.g. University of Minnesota should send in one form for St. Paul campus and one form for Minneapolis campus.

I. PHYSICAL STRUCTURE

    A. Principal S/E library facility

        _____ 1. One principal science/engineering library.

        _____ 2. One principal science/engineering library with independent branches (not reporting to principal S/E library.) List:

        _____ 3. One or more branch/department libraries reporting to principal science/engineering library. List:

        _____ 4. Mixture of 2 or 3 above. List:

_____ 5. More than one principal S/E library, with
or without branch/department libraries.
List (separate principal from branch/
department libraries):

_____ 6. Physically separate S/E library within a
main library facility.

B. Disbursed science/engineering library facilities

_____ 1. Multiple branch/department libraries--no
principal S/E library.

_____ 2. S/E collection completely integrated into
main collection.

C. Other

_____ 1. Other configuration. Please describe.

II. ADMINISTRATION

_____ 1. One administrative head of all S/E libraries.
Title:

_____ 2. Two or more administrative heads. Titles:

_____ 3. Many branch/department library heads reporting
to--

_____ a. Central library

_____ b. Sci/tech coordinator

_____ c. Library heads reporting to individual
academic departments. Titles:

_____ d. Other. Please describe:

_____ 4. No designated administrator for S/E functions.

_____ 5. Other. Please describe:

APPENDIX I (continued)

NAME OF PERSON FILLING OUT FORM_____

TITLE_____

LIBRARY_____

INSTITUTION_____

ADDRESS_____

PHONE NUMBER_____

ELECTRONIC MAIL ADDRESS __BITNET: Node:_____

_____Check here if you would like to receive the results of this
      survey.

SCI-TECH COLLECTIONS

Tony Stankus, Editor

Some problems last longer than others, and certain environmental problems look like they'll last longer than we might. A case in point is radioactive waste management. The positive aspect of this is that many collection managers in sci-tech libraries know that technical information regarding this subject is unlikely to go out of fashion. We can buy with confidence that the material will be of enduring interest. Alas, we do not have budgets with the long half-lives of some of the materials documented within this article. This is where the selections and recommendations of authors Donna E. Cromer and Dena Rae Thomas come in. They help us channel our initial spending wisely, and in addition, give us access to government documents that are relatively low cost, yet likely to be of interest to clients who either have a social or financial concern in the state of radioactive cleanup and storage. Both authors hold the rank of assistant professor at the Centennial Science and Engineering Library, University of New Mexico at Albuquerque, where Donna Cromer is responsible for Physics, Astronomy, and Electrical/Computer Engineering, while Dena Rae Thomas services Civil Engineering Patents, and Government Documents. We thank them for their efforts and commend this paper to your attention.

Radioactive Waste Management and Disposal: Information Sources

Donna E. Cromer
Dena Rae Thomas

INTRODUCTION

Radioactive waste has been a potential problem for at least fifty years, though it is only in the last decade that it has become widely debated. A considerable amount of research has been done, and the results are published in a variety of literature types. The amount of information generated, the range of interests represented, and the numerous document types all combine to make this a particularly tricky area of scientific/technical information. Simultaneously, the demand for information on this subject — from students, scholars, and the general public — is continually escalating. The purpose of this article is to aid librarians in their search for information on radioactive waste management and disposal, both for collection development activities and for reference assistance. Because of the tremendous volume of information available, we make no claim to a comprehensive coverage of the literature.

This article is principally aimed at science/technology librarians, and it primarily reviews the science and engineering literature. Throughout, suggestions are limited to U.S. published items or foreign publications that are easily accessible. A final section provides sources on the social and political aspects of the issues. First is an introduction to the topic, including brief descriptions of the technologies involved. Then examples of the different information sources and literature types are presented with brief characterizations to aid librarians in developing strategies for their use. The specific categories of information covered are: LEGISLATION AND REGULATORY AGENCIES; SUBJECT HEADINGS/KEYWORDS; STA-

TISTICAL INFORMATION; ENCYCLOPEDIAS; BOOKS; CONFERENCE PROCEEDINGS; JOURNALS; TECHNICAL REPORTS; INDEXES/ONLINE AND PRINTED; AND SOCIAL AND POLITICAL PERSPECTIVES.

RADIOACTIVITY

The effective management of radioactive waste is critical because of the potential danger to humans and to the environment. Radioactivity, the spontaneous emission of alpha, beta, or gamma particles from atomic nuclei, can remain for millions of years. It is dangerous because the particles can travel into and damage or destroy living cells by disrupting molecular bonds. Alpha and beta particles cannot travel far, so they are not a great hazard unless they touch the skin, are ingested, or inhaled. Gamma particles are smaller, and can travel greater distances at higher velocities. Cell penetration can occur from a greater distance and there is likelihood of more damage.[1]

When a particle is emitted, the nucleus changes, or decays into what is called a daughter element. This daughter element can either be stable or decay by one or more steps into other elements. "Half-life" is the term used for this rate of decay, and is expressed in years.

A half-life of one year means that it takes one year for half of any amount of a radioactive substance to decay into daughter elements. After one more year, another half decays, so one-fourth of the original amount remains. Half-lives of radioactive elements range from less than a second to billions of years.[2]

RADIOACTIVE WASTE

Radioactive waste occurs in many different forms and originates from the following sources: the nuclear fuel cycle of electric power generation, federal and nuclear weapons research and development, institutions such as universities and hospitals, industrial uses of radioisotopes, and uranium mining and milling. It is typically separated into several categories. High-level radioactive wastes (HLRW) include spent fuel assemblies from nuclear reactors and liquid products from reprocessing spent fuel. These wastes release considerable radioactive decay energy, so heavy shielding is required to protect humans and the environment. Transuranic (TRU) wastes originate from reprocessing spent fuel and from the fabrication of plutonium weapons and reactor fuel. Virtually all of the TRU wastes are products of nuclear defense activities. TRU wastes have an atomic number greater than 92 (uranium) and have half-lives greater than 20 years. Shielding requirements vary. Low-level radioactive wastes (LLRW) are radioactive wastes not classified as any of the others discussed here, generally short-lived and low in radioactivity. They may be cleaning rags and fluids, clothing, tools, glassware, and equipment. LLRW sources are numerous and varied, including medical, government and industrial research, hospital wastes, nuclear power generation, and manufacturing. Uranium mill tailings, the byproducts from uranium mining and processing, are another waste type. A special type of waste that includes several of the other types is the decommissioning of nuclear reactor facilities. Nuclear reactors operate for approximately 30 years, after which time they must be dismantled and disposed of permanently.

Some current and projected data on the amount of wastes to be handled gives an idea of the extent of the problem. At the end of 1987, there were 379,000 cubic meters of HLRW, 3,639,000 cubic meters of LLRW, 250,000 cubic meters of TRU wastes, 116,200,000 cubic meters of uranium mill tailings, and 15,902 MTIHM (a unit of mass rather than volume) of spent fuel assemblies. In the year 2020, it is projected that there will be 9,774,000 cubic meters of LLRW, 335,000 cubic meters of HLRW, and 77,400,000 MTIHM of spent fuel assemblies. From 1976-1982, five nuclear power plants were shut down. It is projected that between the years 1988-2020, 70 reactors will be shut down. The projected volume of wastes from decommissioning is 750,310 cubic meters (*Integrated Data Base for 1988*, pp 15,178,179).

RADIOACTIVE WASTE MANAGEMENT

The complete range of radioactive waste management includes treatment, transport, storage, disposal, and environmental monitoring.

Treatment of radioactive wastes accomplishes three main functions: (1) the concentration of waste into small volumes to minimize the volume of material which must be disposed of, (2) conversion of liquid or gaseous wastes into solid forms which can be incorporated into a glass matrix to drastically reduce chances of reactivity, and (3) the separation of radioisotopes from other wastes for recycling back into the nuclear fuel cycle, also called reprocessing (Barney, 1987, p. 713).

Transportation of radioactive waste poses several problems. The principal hazard is release of radioactivity from damaged shipping canisters. Most of the waste transported for storage, treatment, or disposal is low-level. Canisters used in transporting high-level spent fuel must be able to withstand severe falls, fires, and water immersion without breaking or leaking.

Storage of radioactive wastes is the temporary holding of wastes. Storage time allows for safer handling of the wastes because the radioactivity diminishes as radionuclides decay. Storage and disposal methods must allow for dissipation of waste heat, for the high levels of radioactivity, and for the long life of the radioactivity. In some cases, storage is used as an interim solution when final disposal methods are not yet developed or proven (Barney, 1989, p. 712).

Storage methods vary depending upon the type of waste. Spent fuel assemblies, a form of high-level radioactive waste, are routinely stored in water pools at the reactor site to dissipate the waste heat and to protect workers from high level gamma radiation. Wastes from reprocessed spent fuel are stored in various types of buried steel tanks. Transuranic wastes are a special type of low-level waste. Although their radioactivity level is relatively low, the radionuclides have very long half-lives. Most are currently stored in drums and are kept in vaults or buried trenches at DOE-operated sites for further treatment or disposal. Low-level civilian waste is stored in steel canisters buried in pits at privately operated facilities. Uranium mill tailings are the earthen residues remaining after uranium mining and processing. The chief hazard of mill tailings is radon leakage into the air or groundwater. For many years the waste remained in uncovered piles, but federal cleanup efforts are planned or underway.

Disposal methods proposed through the years range from direct dumping into the oceans, to injection of the wastes into the earth's mantle, to disposal in Antarctic sea ice. "The basic requirement for acceptable final storage or disposal of radioactive waste is the capability to contain and isolate the waste safely until decay has reduced the radioactivity to nonhazardous levels—or at least to levels found in nature" (*McGraw-Hill Concise Encyclopedia of Science & Technology*, 1989, p. 1562).

Currently, the favored type of final repository is a mined-out vault in a geologically stable rock formation. Types of formations considered include basalt, such as at the Hanford Reservation near Richland, Washington; volcanic tuff formations at Yucca Mountain on the Nevada nuclear testing reservation; and salt beds like those in the Palo Duro Basin, Deaf Smith County, Texas. The Waste Isolation Pilot Plant, a complex of deep tunnels in an underground salt bed near Carlsbad, New Mexico, is planned to be a demonstration site for disposal of military transuranic wastes. If no problems appear in the 5-year test period, it is intended to become the permanent repository for that waste type. Yucca Mountain is the proposed site for the final disposal of high-level commercial waste.

Environmental monitoring is an essential activity at any permanent repository. Background levels of radiation must be determined, some measurements must be taken continuously, and others must be taken periodically and then analyzed. At any future permanent repository, DOE and EPA will share the duties of monitoring to detect possible leakage or accidental intrusion.

LEGISLATION AND REGULATORY AGENCIES

Many major pieces of federal legislation direct the control of radioactive materials. The following briefly outlines the history of the most important legislation, and includes date of the original act, public law number and amendments, purpose of the legislation, and the chief agency responsible.

1944—Public Health Services Act (PL 78-410). Wide-ranging legislation which consolidated a number of laws; chiefly relevant because it established research functions. Public Health Service.

1946 — Atomic Energy Act (PL 79-585, many amendments). Established Atomic Energy Commission.

1969 — National Environmental Policy Act (PL 91-190). Among many other provisions, required federal agencies to consider environmental consequences of activities and to submit written reports, i.e., environmental impact statements.

1970 — Clean Air Act (PL 91-604, amended 1977. New bill has passed one House of Congress as of this writing). Set emission standards for air pollutants, including radioactivity, and established air quality standards. Environmental Protection Agency.

1972 — Marine Protection, Research and Sanctuaries Act (PL 92-532). Prohibited ocean disposal of radioactive materials and established marine sanctuaries for preservation and conservation. Environmental Protection Agency.

1974 — Hazardous Materials Transportation Act (PL 93-633). Required establishment of regulations for safe shipment of hazardous materials. Department of Transportation.

1974 — Safe Drinking Water Act (PL 93-523, amended 1977). Established national drinking water standards. Environmental Protection Agency.

1976 — Resource Conservation and Recovery Act (PL 94-580). Defined stringent guidelines for management of solid hazardous wastes. Established restrictions for the disposal of radioactive materials. Environmental Protection Agency.

1976 — Toxic Substances Control Act (PL 94-469). Gave the EPA much greater responsibility for the control of toxic substances. Environmental Protection Agency.

1978 — Uranium Mill Tailings Act (PL 95-604). Established program to clean up uranium wastes, and set strict guidelines on handling and disposal of wastes resulting from uranium processing. Nuclear Regulatory Commission.

1980 — Low-Level Radioactive Waste Policy Act of 1980 (PL 96-573, amended 1985). States given chief responsibility for commercial LLRW disposal. Department of Energy.

1980 — Comprehensive Environmental Response, Compensation and Liability Act (PL 96-510). "Superfund" created to expedite cleanup of hazardous waste sites. Environmental Protection Agency.

1983 — Nuclear Waste Policy Act (PL 97-425). Required that two sites be identified and developed for permanent repositories of radioactive wastes.

Regulatory agencies are also good sources of information. They may have general information desks, public affairs offices, or hotlines to assist the public in obtaining general information or answers to specific concerns. Several of the most pertinent offices are discussed below.

The Environmental Protection Agency is the primary regulatory agency involved in monitoring radioactive waste handling. Their Office of Radiation Programs develops policies for radiation protection, studies measurement and control of radiation, and evaluates new technologies. It also provides technical assistance to states and to other federal agencies, and conducts a national surveillance program that monitors radiation levels.

The Office of Civilian Radioactive Waste Management of the Department of Energy manages the Nuclear Waste Fund. The Office is responsible for overseeing federal programs of storage and disposal of spent fuel and other high-level waste. It also conducts research in the areas of storage and disposal.

The Office of Nuclear Regulatory Research of the Nuclear Regulatory Commission conducts extensive research programs in support of the licensing and regulation of nuclear facilities. Safe disposal of nuclear waste is one of the principal areas of investigation.

SUBJECT HEADINGS/KEYWORDS

The appropriate Library of Congress subject heading is 'Radioactive waste disposal.' There are also more specific terms such as 'Radioactive waste disposal in the ground' and 'Radioactive waste disposal under the seabed.' LC headings are useful because of their widespread use.

Many of the indexing services use 'Radioactive waste management' or 'Radioactive waste disposal' as thesaurus terms. Other common terms in use are 'Nuclear waste' and 'Radwaste.'

STATISTICAL INFORMATION

Specific statistical information such as amounts and types of radioactive waste generated can be very difficult to locate. Two excellent sources that aid in identifying and locating data and one report compiled by DOE are included here.

American Statistics Index. Washington, DC: Congressional Information Service, 1973- .

The correct subject term is 'Radioactive waste and disposal.' Examples of data identified are: projections of spent fuel generation, selected years 1988-2037; shipments of low-level waste to disposal sites, by state; quantity and characteristics of waste generated from commercial reactor decommissioning; and uranium mill tailings inventory data.

Index to International Statistics. Washington, DC: Congressional Information Service, 1983- .

The correct subject term to use is 'Radioactive waste disposal.' Examples of data indexed are: estimated spent fuel and waste generated by reactors in member countries of the International Atomic Energy Agency; and radioactivity problems in the South Pacific.

Integrated Data Base for 1988: Spent Fuel and Radioactive Waste Inventories, Projections, and Characteristics. Prepared by Oak

Ridge National Laboratory for the DOE Office of Civilian Radioactive Waste Management. DOE/RW-0006, Rev. 4, 1988.

This report, updated annually, is replete with data and projections on the amounts of wastes generated in numerous categories, the source of the wastes, and what is currently being done. Also quite useful are definitions of the waste types and a glossary.

ENCYCLOPEDIAS

Encyclopedias are always a good place to start for general background information on any topic. Those listed have good discussions on the subject of radioactive waste management and background material on radioactivity and nuclear power.

Encyclopedia of Physical Science and Technology. Orlando: Academic Press, 1987.

There is good coverage of radioactive waste management in this encyclopedia, at a somewhat technical level. Information is updated in an annual yearbook, first published in 1989.

Energy Deskbook. Oak Ridge, TN: Technical Information Center, US Department of Energy. DOE/IR/05114-1;DE82013966, 1982.

An encyclopedic dictionary, this volume explains technical information at a general level.

McGraw-Hill Encyclopedia of Science and Technology. New York: McGraw-Hill, 1987.

Broad coverage of all areas of science and technology is presented at a less technical level. An annual yearbook exists.

BOOKS

It is impossible to list all the relevant books, so only a few are included. Currency and breadth of coverage are the primary criteria for this collection.

Berlin, Robert E. and Catherine C. Stanton. *Radioactive Waste Management*. New York: Wiley, 1989. 0-471-85792-0.

Chapman, Neil A. and Ian G. McKinley. *The Geological Disposal of Nuclear Waste*. Chichester: Wiley, 1987. 0-471-91249-2.

Krauskopf, Konrad B. *Radioactive Waste Disposal and Geology*. London: Chapman and Hall, 1988. (*Topics in the Earth Sciences*, v.1). 0-412-28630-0.

Lau, Foo-Sun. *Radioactivity and Nuclear Waste Disposal*. New York: Wiley, 1987. (*Research Studies in Nuclear Technology*, v.2). 0-471-91524-6.

Lutze, Werner and Rodney C. Ewing. *Radioactive Waste Forms for the Future*. Amsterdam: North-Holland, 1988. 0-444-87104-7.

Milnes, A.G. *Geology and Radwaste*. London: Academic Press, 1985. 0-12-498070-8.

Roxburgh, I.S. *Geology of High-Level Nuclear Waste Disposal: An Introduction*. London: Chapman, and Hall, 1987. 0-412-29910-0.

CONFERENCE PROCEEDINGS

Many conferences, symposia, and workshops are held each year on the subject of radioactive waste management. Three of the long-lived recurring conferences are described; a sampling of others follows. Entries in this section illustrate the international importance of radioactive waste management and disposal. It is important not to overlook the many other conferences of a more general nature that might include one or two papers or sessions on the topic.

There are a number of ways to find conference proceedings. Standard indexes to conference proceedings include *Proceedings in Print* and *Index of Conference Proceedings*. Most indexing services cover conferences and provide ways to limit to the conference as a whole. Another tactic is to use appropriate subject sub-headings (such as Congresses) when looking in sources such as library catalogs or *Books in Print*. Upcoming conferences are listed in several sources, such as *World Meetings, U.S. and Canada*, and as advertisements or simple listings in journals. See the INDEXES/ON-LINE AND PRINTED section for information on finding individual conference papers.

DOE Low-Level Waste Management Forum (the 10th was held in 1988).

These are sponsored by EG&G in Idaho Falls, ID, in their capacity as DOE contractor for the National Low-Level Radioactive Waste Management Program. A wide range of issues is covered, including regulations, disposal technology, facility development, waste characterization, and performance assessment.

International Symposium on the Scientific Basis for Nuclear Waste Management.

These are published as volumes in the *Materials Research Society Symposium Proceedings*. The most recent one is the XIII, held in 1989 and published in 1990 (*MRS Symp Proc* v176). The scope and participation in these conferences is international, covering a wide range of radioactive waste management concerns.

Symposium on Waste Management. This conference is held every year in Tucson, AZ and is theme oriented. The 1988 conference theme was "Waste processing, transportation, storage and disposal" and the 1985 theme was "Waste isolation in the U.S."

International Conference on Radioactive Waste Management (2nd). Winnepeg, 7-11 Sep 1986. Cosponsored by the American Nuclear Society. Toronto: Canadian Nuclear Society, 1986.

International Symposium on Ceramics in Nuclear Waste Management (3rd). Chicago, 28-30 Apr 1986. D.E. Clark, W.B. White, A.J. Machiels, eds. Held at the 88th Annual Meeting of the American Ceramic Society. Westville, OH: American Ceramic Society, 1986.

International Symposium on Management of Low and Intermediate Level Radioactive Wastes. Stockholm, 16-20 May, 1988. Cosponsored by the Commission of the European Communities. Vienna: International Atomic Energy Agency, 1989.

International Symposium on the Back End of the Nuclear Fuel Cycle. Vienna, 11-15 May 1987. Cosponsored by the Nuclear Energy Agency of the OECD. Vienna: International Atomic Energy Agency, 1987.

International Symposium on the Siting, Design, and Construction of Underground Repositories for Radioactive Wastes. Hannover, 3-7 Mar 1986. Vienna: International Atomic Energy Agency, 1986.

International Waste Management Conference. Hong Kong, 29 Nov-5 Dec 1987. L.C. Oyen, ed. Cosponsored by the International Atomic Energy Agency, the Nuclear Engineering Division of the American Society of Mechanical Engineers, and the American Nuclear Society. New York: American Society of Mechanical Engineers, 1987.

Joint International Waste Management Conference. Kyoto, 22-28 Oct 1989. Cosponsored by the Nuclear Engineering Division of the American Society of Mechanical Engineers, Atomic Energy Society of Japan, the Japan Society of Mechanical Engineers et al. New York: American Society of Mechanical Engineers, 1989.

Natural Analogues in Radioactive Waste Disposal. Brussels, 28-30 Apr 1987. B. C'ome and Neil A. Chapman, eds. London: Graham & Trotman for the Commission of the European Communities, 1987.

Population Exposure from the Nuclear Fuel Cycle. Oak Ridge, 14-18 Sep 1987. Edward L. Alpen, Rowena O. Chester, Darrell R. Fisher, eds. Cosponsored by the American Nuclear Society and Oak Ridge National Laboratory. New York: Gordon & Breach. 1988.

Radioactive Waste Management and Disposal (2nd). Luxembourg, 22-26 Apr 1985. R. Simon, ed. Cambridge: Cambridge University Press. 1986. (*Commission of the European Communities Report* 10163).

JOURNALS

The journal literature is a rich source of information on radioactive waste management. This section lists the more significant journal titles. The first two are devoted almost entirely to the subject of radioactive waste management. The second two concentrate on environmental engineering, while the last four focus on nuclear engineering.

These are only a small number of the potentially useful journal titles. One way to find others is to scan directories of periodicals. The relevant sections in *Ulrich's International Periodicals Directory* are 'Physics – Nuclear Energy' and 'Environmental Studies.' Relevant sections in the *Serials Directory* are 'Engineering, Nuclear Engineering' and 'Sanitary, Environmental Technology.' For access to specific journal articles, see the INDEXES/ONLINE AND PRINTED section.

Radioactive Waste Management and the Nuclear Fuel Cycle
Harwood Academic Pub.; 0739-5876; RWMCD4
Previous title: *Radioactive Waste Management*

Radioactive waste disposal problems created by nuclear power production and the use of radioisotopes in industrial, scientific, and medical fields are reported on. Some issues are proceedings of conferences. Two recent articles are: "Siting a low-level waste disposal facility in California" and "Office of Civilian Radioactive Waste Management ensuring quality assurance in the waste management program."

Waste Management: Nuclear, Chemical, Biological, Municipal
Pergamon Press; 0956-053X; WAMAE2
Previous title: *Nuclear and Chemical Waste Management*

Information on the entire field of hazardous waste (low- and high-level radioactive waste, and chemical and transuranic wastes) is covered in a style that, while technical, is written for the nonspecialist. Included in each issue are book reviews, a list of patents and applications from more than 30 countries, and a software survey section. Two recent articles are: "Final disposal of spent nuclear fuel in Taiwan: a state-of-the-art technical overview" and

"Stochastic analysis of radioactive waste package performance using first-order reliability method."

Environmental Science & Technology
American Chemical Society; 0013-936X; ESTHAG

Legislative action, emerging technology, industrial activity, news, opinions, new products, and technical research articles in environmental engineering are covered. Two recent contributions are: "Mobility of plutonium and americium through a shallow aquifer in a semiarid region" and "The Yucca Mountain project."

Journal of Environmental Sciences
Institute of Environmental Science; 0022-0906; JEVSAG

Articles and reports across a broad range of environmental sciences and technologies are included. One recent contribution is: "Low level radioactive waste disposal – technology and public policy."

JNMM: Journal of the Institute of Nuclear Materials Management
Institute of Nuclear Materials Management; 0893-6188
Previous title: *Nuclear Materials Management: Journal of the Institute of Nuclear Materials Management*

The INMM is a not-for-profit membership organization which publishes on nuclear materials management, safeguards, waste management, transportation, physical protection, and measurements. Two recent articles are: "Yucca Mountain: is it a safe place for isolation of high-level radioactive waste?" and "Spent fuel storage and management in the United Kingdom."

Nuclear Engineering International
Heywood Temple Industrial Pub.; 0029-5507; NEINBF
Previous title: *Nuclear Engineering*

Most issues have one or two articles within broad waste management categories, such as Spent Fuel Transport or High-Level Waste. Two recent articles are: "Developing cask designs in the USSR" and "Getting ready to build the Hanford Waste Vitrification Plant."

Nuclear Safety
Department of Energy/Office of Scientific and Technical
Information; 0029-5604; NUSAAZ

Each issue has a section on Waste and Spent Fuel Management.
It includes brief reports on administrative, regulatory, and technical
activities related to safety aspects of the management of radioactive
wastes and spent nuclear fuel.

Nuclear Technology
American Nuclear Society; 0029-5450; NUTYBB
Previous title: *Nuclear Applications and Technology*

Applications of technology and expertise in the nuclear field are
covered. Most issues have a section on radioactive waste management. One recent article is: "A remote canister-positioning and
glass level detection system.

General science periodicals

General science periodicals occasionally cover radioactive waste
and may be good sources of overviews or background information.
Some of these are: *Bulletin of the Atomic Scientists, New Scientist,
Science,* and *Science News.*

TECHNICAL REPORTS

Technical reports are important sources of information in many
fields of science and technology. They are of particular value in
researching methods of treatment and disposal of radioactive waste
and are highly cited in the literature and indexing sources. Because
energy production and nuclear weapons research are heavily regu-
lated industries, much of the research is federally funded and disseminated in the report literature. The DOE National Laborato-
ries—Argonne, Brookhaven, Lawrence Livermore, Los Alamos,
Oak Ridge, Sandia—as well as other facilities such as the Idaho
National Engineering Laboratory, Pacific Northwest Laboratory,
and the Savannah River Project, are prolific and authoritative producers of reports. Other important agencies that generate reports are
the Environmental Protection Agency, the General Accounting Of-

fice, the Nuclear Regulatory Commission, and the Office of Technology Assessment. For access to the report literature, see the INDEXES/ONLINE AND PRINTED section.

INDEXES/ONLINE AND PRINTED

The list of indexing services appropriate to this topic can be divided into three categories: 'Specialized Indexes,' covering nuclear engineering and energy production; 'Environmental Indexes,' covering diverse areas of environmental control and pollution; and 'Other Indexes.' These indexes either specialize in a particular area of science or technology, or cover the areas broadly. They are listed in brief form only. Most of the indexes have online counterparts for all or part of the dates of coverage of the paper indexes. For more information on the online databases, consult appropriate vendor catalogs, or a database directory.

Specialized Indexes

Energy Research Abstracts. Oak Ridge, TN: Office of Scientific and Technical Information, U.S. Department of Energy, 1976- .

Abstracts and indexes all scientific and technical reports, journal articles, conference papers and proceedings, books, patents, and theses generated by DOE or its contractors. Available online as *DOE/Energy* from 1974 to the present.

INIS Atomindex. Vienna: International Nuclear Information System, 1970- .

Atomindex is the major abstracting service in the field of nuclear engineering. Its stated purpose is to identify publications relating to the peaceful uses of nuclear science. It is a cooperatively produced database, international in scope, and covers a variety of literature types. The file is available online from several vendors from 1976.

Nuclear Science Abstracts. Washington, D.C.: U.S. Energy Research and Development Administration, 1948-1976. (Published by the U.S. Atomic Energy Commission prior to Feb. 1975.)

Abstracts and indexes technical reports sponsored by AEC and other U.S. government agencies. Covers books, conference papers, patents, and the journal literature worldwide. Superseded in part by *INIS Atomindex* and *Energy Research Abstracts*. Online, the file includes material from 1948-1976.

Environmental Indexes

Environment Abstracts. New York: Bowker A&I Publications, 1974- .

Environment Abstracts is the major abstracting service for the broad range of environmental topics. Covers journal articles, reports, government documents, conferences and symposia, patents, and newspaper articles. Available online through a number of vendors as *ENVIROLINE* from 1971 to the present.

EPA Publications Bibliography: Quarterly Abstracts Bulletin. Springfield, VA: National Technical Information Service, 1977- .

Indexes EPA technical reports. Access points include author, title, subject, keyword, and others. The *EPA Cumulative Bibliography, 1970-1976*, provides retrospective coverage.

Pollution Abstracts. Bethesda, MD: Cambridge Scientific Abstracts, 1970- .

Comprehensive coverage of all aspects of pollution. A variety of literature types are included: journal articles, conference proceedings and papers, books, government documents, and patents. Available online through several vendors from 1970 to the present.

Other Indexes

Applied Science & Technology Index and *General Science Index*. New York: H.W. Wilson. Available online only on Wilsonline, *Applied Science & Technology* from October 1983 and *General Science* from May 1984.

Biological Abstracts. Philadelphia, PA: BioSciences Information Service, Biological Abstracts. Available online as *BIOSIS*, 1969 to the present.

BIOSIS — see *Biological Abstracts*.

Chemical Abstracts. Columbus, Ohio: Chemical Abstracts. Available online through many vendors, although the abstracts are available exclusively on STN. 1967 to the present.

COMPENDEX — see *Engineering Index*.

Engineering Index. New York: Engineering Information. Available online as *COMPENDEX* through a variety of vendors, from 1970 to the present.

Government Reports Announcements and Index. Springfield, VA: National Technical Information Service. Online as NTIS from a number of vendors, 1964 to the present.

INSPEC — see *Physics Abstracts*.

NTIS — see *Government Reports Announcements and Index*.

Physics Abstracts. [London]: IEE Available online from several vendors as part of the *INSPEC* file, 1969 to the present.

SCISEARCH — see *Science Citation Index*.

Science Citation Index. Philadelphia: Institute for Scientific Information. Online as *SCISEARCH*. 1974 to the present.

SOCIAL AND POLITICAL PERSPECTIVES

Because of the enormity of the problem and the amount of media attention, radioactive waste management interests a surprising variety of people. Information seekers may include attorneys, engineers, environmentalists, ministers, scientists, students, and teachers, to name a few. This section lists sources presenting social and political perspectives, as well as several of the more well-known special interest groups.

Burns, Michael E., ed. *Low-level Radioactive Waste Regulation: Science, Politics, and Fear*. Chelsea, MI: Lewis Publishers, 1988. 0-8737-1026-6.

Jacob, Gerald. *Site Unseen: The Politics of Siting a Nuclear Waste Repository.* Pitt Series in Policy & Institutional Studies. Pittsburgh, PA: University of Pittsburgh Press, 1990. 0-8229-3640-2.

Social and Economic Aspects of Radioactive Waste Disposal: Considerations for Institutional Management. Panel on Social and Economic Aspects of Radioactive Waste Management, Board on Radioactive Waste Management, National Research Council. Washington, DC: National Academy Press, 1984. 0-3090-3444-2.

Vance, Mary A. *Radioactive Waste Disposal: A Bibliography.* Monticello, IL: Vance Bibliographies, 1985. 0-8902-8386-9.

A great number of indexes carry references to articles expressing social and political concerns. A few which may be useful include: *Index to U.S. Government Periodicals, Magazine Index, National Newspaper Index, Public Affairs Information Service, Social Planning, Policy & Development Abstracts*, and *Sociological Abstracts*.

Radioactive waste management is a major concern for environmental and public interest groups. The following list of organizations is a sampling of the many active groups: Environmental Action, Environmental Defense Fund, Greenpeace International, Natural Resources Defense Council, Public Citizen, Radioactive Waste Campaign, Sierra Club, and Wilderness Society. Local chapters of these or other organizations may be identified in your local Yellow Pages under the heading 'Environmental, Conservation, and Ecological Organizations.' National offices are available from a number of sources, including the *Encyclopedia of Associations*.

CONCLUSION

Radioactive waste management is a major social and technological problem. The information sources covered in this article are extensive, illustrating how pervasive the subject has become. It is not intended to be a comprehensive compilation, which would be impossible in a document of this length. Certain useful categories of information have been omitted, such as government publications, bibliographies, and professional organizations, and some descriptions are condensed rather than complete. We think that the information provided will be useful to any information professional interested in the subject of radioactive waste management.

NOTES

1. Alpha particles consist of two protons and two neutrons (identical to a helium nucleus). Their range is 1.2 to 3.2 inches in air. Beta particles are identical to electrons, and their range is several feet in air. Gamma particles are identical to high-energy X-rays, with no mass or electric charge and can travel considerable distances, varying according to the original energy state (Glasstone, 1982, p.309).

2. One important decay sequence is that of uranium 238, which is the most abundant isotope of uranium. U 238 decays by the emission of an alpha particle into thorium 234 and then by emission of a beta particle into proactinium 234. Another daughter element is radon 226, which decays through a number of steps into Ra 222, polonium, bismuth, and thalium. The half-life of U 238 is 4,490,000,000 (4.49 x 10-9) years (Glasstone, 1982, p.308-9; Eisenbud, 1989, p.80).

REFERENCES

Barney, G S. "Radioactive Wastes." *Encyclopedia of Physical Science and Technology.* New York: Academic Press, 1987, v.11, pp. 708-722.

Eisenbud, Merril. "Environmental Radioactivity." *Encyclopedia of Physical Science and Technology, 1989 Yearbook.* New York: Academic Press, 1989, pp. 75-90.

Glasstone, Samuel. *Energy Deskbook.* Oak Ridge, TN: US DOE Technical Information Center. DOE/IR/05114-1; DE82013966, 1982.

Integrated Data Base for 1988: Spent Fuel and Radioactive Waste Inventories, Projections, and Characteristics. Prepared by Oak Ridge National Laboratory for U.S. DOE Office of Civilian Radioactive Waste Management. DOE/RW-0006, Rev. 4, 1988.

Lombardo, Thomas G. "Nuclear Waste Disposal: Where and How?" *IEEE Spectrum.* v17, n12, pp. 38-43, 1980.

McGraw-Hill Concise Encyclopedia of Science & Technology. Sybil S. Parker, ed. New York: McGraw-Hill, 1989. 2nd edition.

NEW REFERENCE WORKS
IN SCIENCE AND TECHNOLOGY

Arleen N. Somerville, Editor

Reviewers for this issue are: Laura Delaney (LD), New York Public Library; Isabel Kaplan (IK), University of Rochester; Richard Kaplan (RK), Rensselaer Polytechnic Institute; Kathleen Kehoe (KMK), Columbia University; Donna Lee (DL), University of Vermont; Robert Rittenhouse (RJR), University of Akron; Arleen Somerville (ANS), University of Rochester.

EARTH SCIENCES

Encyclopedia of solid earth geophysics. Edited by David E. James. New York: Van Nostrand; 1989. 1328p. $125. ISBN 0-442-24366-9. (Encyclopedia of Earth Sciences Series)

> This is another in the valuable Encyclopedia of Earth Sciences Series. The entries emphasize what the subject is about, why it is important, and how it contributes to a broader understanding of the earth. Illustrations, graphs, and charts are used liberally. Each entry includes a bibliography with references of the most important sources from the past twenty years. Cross references appear with each entry, with an overall subject index. No other encyclopedia covers geophysics comprehensively. The many entries under earthquakes and seismicity attest to this encyclopedia's extensive coverage. Essential for all collections that serve clientele conducting research or writing papers in geophysics. (ANS)

The geophysical directory. 45th ed. Edited by Claudia LaCalli. Houston: The Geophysical Directory, Inc.; 1990. 528p. $35. ISBN not available.

Now in its 45th edition, this comprehensive directory continues to supply a wealth of information in the field of geophysical exploration. It lists approximately 3,600 companies providing geophysical equipment, supplies or services as well as mining and petroleum companies utilizing geophysical techniques. Entries on individual companies include such standard information as company name, address, and telephone number along with names of principal executives and sales personnel.

Special features of note include a useful list of geophysicists and geologists who direct geophysical operations as well as a list of companies supplying geophysical software. A separate section describes the activities of national and international geophysical societies. Completing the text is an extensive company index. An excellent reference guide for technical collections supporting geophysical exploration. (LD)

Guide to U.S. map resources. 2nd ed. Compiled by David A. Cobb. Chicago: American Library Association; 1990. 495p. $25. ISBN 0-8389-0547-0.

Although this edition comes only four years after the first edition, this *Guide* expands the number of collections included and the information provided. The extensive profiles of each map collection include the following information: responsible person; employees; area and special strengths; special collections; holdings by format and chronological coverage;; percent of maps cataloged and classified; classification systems and utilities; preservation methods; square footage; equipment; publications; hours; online search systems; circulation and use data; interlibrary loan services; and phone, fax, and email numbers. The newly added indexes of responsible individuals, collection strengths and institution names will be valuable resources. This guide provides more information about each collection than does the 1985 *Map Collections in the United States and Canada* (4th ed., New York, Special Libraries Association; 1985), but the SLA guide includes some U.S. collections, as well as Canadian collections, not found in this guide. An essential title for all public and academic library collections and industrial libraries whose clientele require map information. (ANS)

Rocks, minerals & fossils of the world. By Chris Pellant. Boston: Little, Brown and Company; 1990. 175p. $17.95. ISBN 0-316-69796-6.

Written for amateur collectors, this richly illustrated guide aims to assist the reader in identifying a wide variety of rocks, fossils, and minerals. The text, divided into three main sections, examines how rocks, minerals, and fossils are formed and how they can be located, identified, and collected.

Section one addresses igneous, metamorphic, and sedimentary rocks respectively and includes numerous field photographs to illustrate the text. Section two, devoted entirely to minerals, provides useful data on each specimen's physical properties and mode of occurrence. The final section examines a variety of fossils including corals, sponges and algae, plants, molluscs, brachiopods, echinoderms, arthropods, fish, teeth, vertebrates, and trace fossils.

Separate rock, mineral, and fossil indices are included as well as a short glossary and brief list of further readings which should prove helpful to first-time collectors. Although too elementary for research level collections, this text should prove useful in public, high school, and college libraries. The inexpensive price tag also makes it an excellent candidate for personal purchase. (LD)

ENGINEERING AND TECHNOLOGY

Eshbach's handbook of engineering fundamentals. 4th ed. Edited by Byron D. Tapley. New York: John Wiley and Sons; 1989. 2368p. $74.95. ISBN 0-471-89084-7.

The *Handbook*'s primary objective is to bring together the fundamentals of engineering's various disciplines — mechanical, electrical, chemical, civil — in a way that will serve the needs of students, practicing engineers, and engineering managers. It is a wealth of tables, charts, diagrams, equations, definitions, and explanations.

In recognition of the considerable changes in engineering techniques and applications since the previous (1975) edition, it has been significantly revised. The Preface notes that there are fewer mathematical and trigonometric tables due to the widespread use of handheld calculators/computers and desktop computers to produce these kinds of data. A chapter has been added on computer science and the chapter on automatic control has been greatly expanded. There is additional emphasis on astronautics and aeronautics (separate chapters in this edition) in light of the important role of space exploration and application, and substantial updates on electromagnetics, electronics, acoustics, light, engineering economics, finite element method and differential equations. This edition uses international standard units throughout. Bibliographies have been updated, too, but there are few references to works published after 1980. Nevertheless, the *Handbook* ranks as a core collection purchase. Recommended for research and technical libraries serving engineers. (IK)

Fiber optics standard dictionary. 2nd ed. By Martin H. Weik. New York: Van Nostrand Reinhold; 1989. 352p. Paper text ed. $32.95. IBN 0-442-23387-6.

This new edition, which updates the 1981 *Fiber Optics and Lightwave Communications Standard Dictionary*, is compiled by an expert who actively participates in establishing fiber optics standards. While a number of organizations are currently preparing terminology definitions and standards, most are aimed at specific audiences and therefore include terms limited to that specialty. This comprehensive dictionary of several thousand terms goes beyond the scope and content of published and pending definitions. Explanatory material, examples, illustrations, and cross references are used liberally. A twelve-page bibliography is included. An essential dictionary for all libraries serving individuals who use fiber optics. (ANS)

Handbook of organic coatings, a comprehensive guide for the coatings industry. Edited by Raymond B. Seymour and Herman F. Mark. New York: Elsevier; 1990. 500p. $85. ISBN 0-444-01519-1.

What a great disappointment this book is. Despite its title, it would be of little use to persons actively involved in the coatings industry. Topics range from "History of paints" (Chapter 1) to "The modern coatings industry" (Chapter 30). In between are chapters on types of materials, such as acrylic polymers, amino resins, epoxies, fluorocarbons, rubber resins, urethane coatings; and related topics including application techniques, extenders and other additives, and testing of coatings. But chapters are short and not nearly as comprehensive as in the *Encyclopedia of polymer science and engineering*. And although this a handbook and not an encyclopedia, it is not as fact-filled as one would expect from a technical handbook. The chapter on epoxies (3 pages) includes two sentences under the sub-heading "Water-borne epoxy coatings." Chapter bibliographies are also short. There are 18 references in the chapter on alkyds; *EPSE* has 103. Each chapter includes a glossary (a bit redundant) and many terms are very elementary. It is necessary to define TiO_2 or E.P.A. or viscosity? There are also irritating errors in punctuation and text spacing. Not recommended. (IK)

Inventing and patenting sourcebook: how to sell and protect your ideas. 1st ed. By Richard C. Levy. Edited by Robert J. Huffman. New York: Gale Research; 1990. 922p. $75. ISBN 0-8103-4871-3.

With this text the author, an experienced inventor himself, succeeds in providing a comprehensive, up-to-date sourcebook on how to develop, patent, license, and market an idea or concept. Along with plenty of "how-to" advice on protecting and licensing an invention, this text supplies a wealth of directory-type information. Names and addresses are supplied for a variety of patent-related individuals, groups, and institutions including invention consultants and research firms, national and regional inventor associations, venture capitalists, patent depository and special libraries, and registered patent attorneys and agents.

Other helpful information includes a chapter on federal funding sources and a separate list of online resources for inventors. Completing the text are two useful appendices the first of which provides the entire index to the U.S. patent classification system. Appendix B, in turn, is a reprint of the official directory of the Patent and Trademark Office. A substantial master index facilitates access to the directory section. An excellent, one-stop reference source for technical collections of all levels. (LD)

New York State annual energy review: energy consumption, supply, and price statistics, 1970-1988. Albany, NY: New York State Energy Office; 1989. 97p.

With the increased emphasis on energy supply and consumption, cumulative data by state is valuable for scientific, economic, and political clientele in many libraries. Data, presented in graphs and charts, cover: consumption by fuel, sector (residential, commercial), and utility; prices by sector; costs by fuel and sector; and sources of energy supplies for New York State. A final section compares New York State consumption to that of the United States overall. (ANS)

Practical electrical wiring: residential, farm, and industrial. 15th ed. By H.P. Richter and W. Creighton Schwan. New York: McGraw-Hill; 1990. 645p. $32.95. ISBN 0-07-052393-2.

Updated to meet the requirements of the 1990 National Electrical Code (NEC), the 15th edition of this best-selling handbook covers electrical wiring for houses, farms, and small commercial projects. Addressed are a variety of topics ranging from how to lay out and plan an electrical wiring system to wiring for special appliances and modernizing outdated wiring. Commonly used wiring methods such as overcurrent protection, grounding, and installation are also covered. (Note: While the methods discussed herein are in accordance with the 1990 NEC, the authors do not attempt to supply a detailed explanation of the Code.)

In addition to updating information from previous editions, the text also explores brand new topics including compact fluorescent lamps and nonincendiary circuits for hazardous locations. Less technical and detailed than *American Electricians' Handbook* (McGraw-Hill, 1987), this text is suitable for novices as well as professional electricians. A substantial index as well as an abbreviated bibliography and glossary of terms are included. An accessible, practical guide for technical collections and a handy home reference for the fix-it-yourselfer. (LD)

The software encyclopedia 1990. By R.R. Bowker's Database Publishing Group and Systems Development Department. N.Y.: Bowker; 1990. 2 vols. 2522p. $189.95. ISBN 0-8352-2762-6.

The Software Encyclopedia is one of the best print compilations of microcomputer software covering all subject areas including engineering, science and medicine. It does not contain information on software designed for schools which can be found in Bowker's companion edition called: *Software for Schools*.

The *Software Encyclopedia* is divided into 2 volumes. Volume 1 contains the Title Index which includes hardware requirements, price, publisher, annotation; and the Publisher Index which lists each publisher, address, phone number and products. Volume 2 is arranged by system hardware such as Macintosh, Atari, IBM, Unix, etc. Within each system, the software is arranged by subject and specific application.

I find it to be very easy to use and fairly complete given the fact that new microcomputer software is being written daily. The publisher index is particularly useful in finding addresses of new and sometimes very small (1 or 2 person) software firms.

The only concern that I have with this otherwise excellent reference book is the lack of very specific subject headings in science and engineering. While I recommend the *Software Encyclopedia* to all academic libraries, Engineering Information, Inc. produces a software guide with a very detailed subject index called: *Engineering & Industrial Software Directory*. It is published every 2 years for around 85 dollars. If your library is primarily engineering and on a limited budget, then the *Engineering & Industrial Software Directory* should suffice. Large academic science/engineering libraries should, in my opinion, purchase both as well as consider the purchase of the *Faulkner Report on Microcomputers and Software* ($564 annually) or the *Datapro Directory of Microcomputer Software* ($725 annually). Both are updated monthly but are weak in sci/tech software and fairly expensive.

When considering the purchase of software directories, keep in mind that the most up-to-date information is found in a number of online databases such as the *Software Directory* (Dialog file 263), *Buyer's Guide to Micro Software* (Dialog file 237), or the *Microcomputer Software Guide* (Dialog file 278. This is the online equivalent of the *Software Encyclopedia*). (RK)

Television and audio handbook for technicians and engineers. By K. Blair Benson and Jerry C. Whitaker. New York: McGraw-Hill; 1990. Various pagination. $39.95. ISBN 0-07-004787-1.

Written for service technicians, operators, and staff engineers, this practical handbook covers the daily maintenance and operation of television and audio equipment. Much of the text and many of the illustrations are drawn from two other McGraw-Hill reference books: the *Television Engineering Handbook* and the *Audio Engineering Handbook*. (The authors feel that the increasingly close relationship between television and audio technologies warrants such a combined reference source as this one.)

In addition to examining fundamental concepts of television and audio circuit design and performance, the text also addresses systems and component operation, maintenance, and testing. Among the many topics covered are video-tape recording and reproduction, electronic editing, satellite communications systems, interfacing audio systems, elements of receiver design, and magnetic-tape manufacture. A subject index and limited chapter references are included. Suitable for public libraries and technical colleges. Should also prove a useful on-the-job reference source. (LD)

HEALTH SCIENCES

Allied health education directory 1990. 18th ed. Chicago: AMA; 1990. 344p. $25.95. ISBN 0-89970-377-1.

The AMA's Committee on Allied Health Education and Accreditation (CA-HEA) approves educational programs in 26 allied health disciplines including, among the more unusual, electroneurodiagnostic technology, medical illustration, perfusion, and sonography. Entries list an address, phone, program director, class size, program length, tuition, and hospital affiliations. Much of the book is given over to a statistical profile of the CAHEA-accredited programs. (DL)

The consumer health information sourcebook. 3rd. ed. By Alan M. Rees and Catherine Hoffman. Phoenix: Oryx; 1990. 210p. $39.50. ISBN 0-89774-408-X.

The authors state in the preface that this guide covers 750 books, 79 magazines and newsletters, over 700 pamphlets, 30 health information clearinghouses, 103 toll-free hotlines, 175 organizations and 49 professional publications of use to health care consumers. Most chapters are devoted to different sources of information, such as clearinghouses and hotlines, health magazines and newsletters, books, pamphlets, and professional literature. Within the books and pamphlets chapters, the sources are grouped by topic, such as AIDS, allergies, Alzheimer's disease, diabetes, health of children, nutrition, physical fitness, and pregnancy and childbirth. Each source of information is cited along with its price and a concise, but fairly thorough review of between 100 and 300 words. Reviews indicate intended use and audience. Indexes give access by author and title of publication, as well as by subject. Very helpful for libraries building a health sciences collection for the general public, and perhaps for nurses and physicians recommending publications for their patients. (DL)

Dictionary of medical syndromes. 3rd ed. By Sergio I. Magalini, Sabina C. Magalini, and Giovanni de Francisci. Philadelphia: Lippincott; 1990. 1042p. $79.50. ISBN 0-397-50882-4.

Considerably more helpful than the standard medical dictionary, yet far more succinct than textbooks, this publication is one of a kind. For each disease the authors explain the symptoms, the diagnosis, the cause, the course of the disease, treatment, and prognosis, with just about as few words as possible. Each entry is accompanied by a bibliography. While the citations are not as up-to-date as one might hope, the authors intention is to list the most important papers, not the most recent. Bibliographies begin by citing the first article to describe the syndrome, and follow with any papers that contain a major revision of our understanding of that syndrome. The index is everything it should be. The authors note that a large proportion of the changes for this revision have been in the field of genetics. Every library with any need for medical reference books should own this one. (DL)

Health and mental health: directory of key legislators and legislative staff. Denver: Health and Mental Health Program, National Conference of State Legislatures; 1989. 83p. $10. ISBN 1-55516-686-5.

This directory lists the names and phone numbers of state representatives and senators who serve on health care related committees within their state governments. Lobbyists and others interested in health care policy may find this publication supplements the *National Health Directory*, which provides the same information for state agencies involved in health care issues. (DL)

Medical quotes: a thematic dictionary. Edited by John Daintith and Amanda Isaacs. New York: Facts on File: 1989. 260p. $30. ISBN 0-8160-2094-9.

Collections of quotations are always fun to read, and this one is no different. Some of these aphorisms may be found in general compilations, but many are unique. While the book's subject arrangement may not please everyone, it is more useful than a chronological or "author" arrangement for the purposes of lecturers and speech writers searching for a quote to lighten or enlighten their talks. Keyword and name indexes are provided. (DL)

Prevention 89/90: federal programs and progress. Washington, DC: US Department of Health and Human Services, Public Health Service, Office of Disease Prevention and Health Promotion; 1990. 192p. $13. No ISBN given.

Chapter one briefly describes Public Health Service programs which disseminate health information. Chapter two uses charts and tables to report on the current health status of the nation. Chapter three outlines health promotion services and research sponsored by other government agencies and by private organizations in cooperation with the federal government. The volume concludes with fifty-five pages of tables detailing the funding levels for each program in FY87 and FY88. Your tax dollars at work. (DL)

Relative values for physicians. 3rd ed. Relative Value Studies, Inc. New York: McGraw Hill; 1989. 1 vol. Looseleaf; approx. 470p. $189. ISBN 0-07-600550-X.

The relative values in question here are monetary, not ethical. The corporation responsible for this publication has analyzed thousands of physician procedures to determine which require more time, skill, and risk, and therefore deserve greater compensation. The book is meant to be used by physicians and insurance companies as a guide for setting fees. (DL)

The universal healthcare almanac. Phoenix: Silver & Cherner, R-C Publications; 1990. 1 vol. Looseleaf. $125. ISBN 0-93-331102-8.

Between *The Statistical Abstract of the United States* and American Hospital Association and the Health Care Financing Administration publications, medical school libraries already have most of this data somewhere in their reference collection. The Almanac does perform some unique calculations and comparisons of the data from these other sources. For example, the section on hospitals uses AHA and US Bureau of the Census information to assess the number of hospital beds, admissions and services per 1,000 people in the years from 1981 to 1989. In addition diagnostic statistics, Medicare/Medicaid admissions figures, and other heretofore unpublished government data can be found here. (DL)

LIFE SCIENCES

The ants. By Bert Hoelldobler and Edward O. Wilson. Cambridge, MA: Belknap Press/Harvard University Press; 1990. 732p. $65. ISBN 0-674040-75-9.

The joint efforts of two distinguished life scientists, Bert Hoelldobler and Edward O. Wilson, have produced an impressive book on the ants. This is a comprehensive treatise which summarizes the current understanding of the anatomy, physiology, ecology, social organization, behavior, and evolution of all 292 known genera of ants. The volume is illustrated by a plethora of line drawings, paintings and photographs of the insects. The authors have included taxonomic schema, data tables, geographical distribution keys, a glossary, a 55 page bibliography, and an index in the volume. The book is extremely well written, and it will be understood by most nonscientists. I recommend *The Ants* for all biology and life sciences collections. It will be of particular interest to biologists, sociobiologists, ethologists, and ecologists. (KMK)

Swallows and martins: an identification guide and handbook. By Angela Turner and Chris Rose. Boston: Houghton Mifflin; 1989. 258p. $35. ISBN 0-395-51174-7.

This bird identification guide to swallows and martins is a high quality in-depth field manual of the songbird family, Hirundinidae. Being a new major book on the swallows distinguishes this book in the field of ornithology. This title is an exciting addition to the literature on the songbirds, and is international in scope. The introductory sections include: morphology, feather plumage, evolution, classification, food, behavior, breeding, migrations, populations, and environmental influences.

The book is structured into bird family characteristics, drawings, descriptions of 74 species, bibliography, and index. While the shapes and detailing of feather plumage are excellent, the colors are not quite correct such as with barn swallows and purple martins. Discussions of race, while useful, do not sufficiently allow for questionable field data, faded museum specimens, and interbreeding. Species descriptions summarize field characters, habitat, populations, food, behavior, including social organizational and sociosexual, song, measurements, and races. Colored plates indicate species occurrence in the world. Recommended for major academic and public libraries, special libraries, and museums in the natural sciences. (RJR)

PHYSICAL SCIENCES

Acronyms and abbreviations in molecular spectroscopy; an encyclopedic dictionary By Detlef A.W. Wendisch. Berlin, NY: Springer-Verlag; 1990. $60. ISBN 0-387513-48-5.

This dictionary focusses on acronyms and abbreviations used in nuclear magnetic resonance, infrared, Raman, electron spin resonance, and other molecular spectroscopic methods. This is the most extensive listing of spectroscopic acronyms. Definitions vary from one-fifth of a page to two pages with text in readable but more compressed format than Homans' book (*Dictionary of concepts in NMR* — see review in this issue). Graphics rarely appear; an occasional equation is included. All entries have at least one reference for further reading and many have up to five or even ten references. An important dictionary for collections serving researchers who use these common spectroscopic methods. (ANS)

Analytical instrumentation handbook. Edited by Galen Wood Ewing. New York: Marcel Dekker; 1990. 1071p. $195. ISBN 0-8247-8184-8 (alkaline paper).

This handbook describes in considerable detail the major techniques of spectrochemical, electrochemical, and chromotographic analysis, as well as selected other techniques. The chapters include extensive description of theory and practical applications, commercially available instruments, use of the instruments, and current trends. New extensions of techniques, such as two-dimensional nuclear magnetic resonance, are included with references to primary articles. Each chapter concludes with an extensive bibliography. Because each chapter is so inclusive (nuclear magnetic resonance chapter is 73 pages) and broad based, this handbook can serve as a one-volume resource in place of a number of monographs. Useful for libraries serving researchers and full-time technicians. (ANS)

CRC handbook of chemistry and physics. 71st ed. David R. Lide, Editor-in-Chief. Boca Raton, FL: CRC Press; 1990. $95. ISBN 0-8493-0471-7.

With the 71st edition of this classic reference book, the new editor initiated the first of a series of changes that will revise and expand tables, add new features, and reorganize the contents. About 20% of the approximately 300 tables are completely new or significantly updated. These tables include, for example, properties of superconducting materials, properties of common amino acids, table of isotopes, diffusion in semiconductors, and vapor pressures of several categories of substances. Increased coverage of health and safety related information includes: handling and disposal of chemicals, threshold limit values for airborne contaminants, properties of photochemical data for atmospheric chemistry. The task of converting to SI units and modern symbols and terminology was completed. The reliable sections were retained, such as: properties of the elements and inorganic compounds; thermodynamics, electrochemistry, kinetics; fluid properties; biochemistry and nutrition; analytical chemistry; molecular structure and spectroscopy; atomic, molecular, and optical physics; nuclear and particle physics; properties of solids; geophysics; practical laboratory data; and mathematical tables. (ANS)

Chemical technicians' ready reference handbook. 3rd ed. By Gershon J. Shugar and Jack T. Ballinger. New York: McGraw-Hill; 1990. $75. ISBN 0-07-057183-X.

This handbook does an excellent job of explaining frequently used laboratory techniques in easy to understand terms and in step-by-step detail. Basic techniques include, for example, recrystallization, filtration, extraction, distillation, pH measurement, gas chromatography, and measurement of more than 15 physical properties. New chapters cover spectroscopy techniques, such as nuclear magnetic resonance, infrared, emission, atomic absorption, visible, and ultraviolet. New sections also emphasize laboratory safety: laboratory safety rules, safe handling of chemicals, laboratory first aid, selected sources of substance toxicity information, and waste disposal of chemicals. Clear graphics are used liberally. References send readers to major references. Because the emphasis is on step-by-step procedures, the chemist will find it useful to consult other publications, as those listed in the chapter bibliographies or publications such as *The Chemist's Companion*, for data needed to interpreting the experimental results. An important handbook for all libraries which serve chemists. (ANS)

Dictionary of colloid and surface science. Paul Becher. New York: Marcel Dekker; 1990. 202p. $80. ISBN 0-8247-8326-3 (alkaline paper).

This dictionary provides brief definitions of terms used by researchers in colloid and surface science. Definitions range from two lines to one-third of a page. Many of these terms are not found in general chemical dictionaries. A special feature is inclusion of names of chemists, with brief biographic information, who have contributed to the fields over the last two hundred years. A useful book for libraries with strong colloid and surface science collections. (ANS)

Dictionary of concepts in NMR. By S.W. Homans. Oxford: Oxford University Press/Clarendon Press; 1989. 343p. $80. ISBN 019-855274-2.

This dictionary explains technical jargon and some acronyms used in nuclear magnetic resonance. The explanations are extensive — range between one-half page for axial peak to four and a half pages for two-dimensional NMR. Graphics appear often, as do math equations. Many entries include references for further reading. Cross references are liberally sprinkled in the alphabetic arrangement. An indispensible dictionary for all collections serving researchers who use NMR. (ANS)

SCI-TECH IN REVIEW

Karla Pearce, Editor

CD-ROMS IN EUROPE AND THE US

Chen, Ching-Chi; Raitt, David. Optical products in American libraries and information centers: similarities, differences, and trends. *Microcomputers for Information Management*. An International Journal for Library and Information Services. 7 (1): 1-23; 1990 March.

This survey compares the use of optical disks, in particular CD-ROMs, in academic, public and special libraries in the US, with their use in European libraries. Besides the obvious—more CD-ROMs were available in 1988 than in 1987, more were purchased by libraries in the latter year, and smaller libraries had, on the average fewer CD-ROMs than larger ones—some less intuitive data are reported. Cost was mentioned more often as a problem for the European libraries, whereas difficulty in use was more of an issue for US libraries. In response to questions about user demand for CD-ROMs, 25% of European libraries reported that there was none, while another 20% professed no interest in the technology. In those that were interested, Medline and Books in Print Plus were most popular in Europe, unlike the US where Infotrac and Eric were most heavily used. One suspects that use in Europe may be limited to research libraries, whereas they are used more widely, in public and

151

school libraries as well, in the US. The authors conclude that the Europeans lag 2-3 years behind US libraries in their use of optical technology. (KJP)

USE OF THE OLDER PHYSICS LITERATURE

Gupta, Usha. Obsolescence of physics literature: exponential decrease of the density of citations to Physical Review articles with age. *Journal of the American Society for Information Science.* 41(4): 282-87; 1990 June.

Conventional wisdom holds that the use of scientific literature decreases as it ages. This article supports that intuition by showing that citation densities—the number of citations to articles in a particular year, divided by actual numbers of articles published in that year—will decrease exponentially, with age. Fifteen leading physics journals were surveyed for their citations to articles in Physical Review. The half-life of those articles—the time during which one-half of the currently active literature was published—went back just 4.9 years. One not surprising exception is referenced: citations to heavily cited articles showed a sharp increase in citations at first, then declined gradually. (KJP)

DIALOG VS. CHEMICAL ABSTRACTS

Hogan, Tom. Dialog sues ACS over access to data. Is it possible — the two most influential organizations in the online database industry at opposite ends of a law suit? *Information Today.* 7 (7): 1, 5-6; 1990 July-August.

In June of 1990, Dialog filed an anti-trust suit against the American Chemical Society on the grounds that ACS would not allow Dialog full access to their database, despite the more than $15 million in federal funding that went into its creation. Dialog claims that Chemical Abstracts' 1988, 60% share of the chemical information marketplace, an increase of 15% since 1983, is leading toward a monopoly. CA had always limited access to their abstracts to users of their own database service, CAS Online, but when they recently announced their intention to remove connection table data, which is essential to graphical representation of molecular structures, Dialog finally chose to fight back. CA's unofficial reply to Dialog's suit has been that their database is available and accessible to the entire scientific world and that the high quality of this service is the source of their strong market share. Users of chemical information are sure to be affected by the outcome of this suit. (KJP)

BE PERSISTENT, PATIENT AND POSITIVE

Hulbert, Doris. Assertive management in libraries. *The Journal of Academic Librarianship*. 16 (3): 158-162; 1990 July.

After describing typical outcomes that may result from passive or aggressive management, the author describes some of the happier results that may be enhanced by assertive management techniques. Effective communication, including careful listening and clear explanations of issues, and an ability to compromise and negotiate will help managers to solve personnel and other problems. Persistence, patience, positive recognition, and effective criticism are needed to encourage the best work performance. A quick and useful approach to effective library management. (KJP)

TECHNOLOGICALLY NAIVE ADMINISTRATORS

Klobas, Jane E.; Clyde, Laurel A. Hit or miss: how librarians find out about new technology. *The Electronic Library*. 8 (3): 172-180; 1990 June.

Although almost all librarians who responded to this survey, taken in Western Australia in 1989, agreed on the importance of keeping up with new technology, few appeared to be making any substantial effort to maintain this awareness. They chose rather to consult readily available sources such as popular journals. But even most of that minority who did make the effort were not satisfied with the result. A nationwide online service to monitor new technical developments was recommended as a solution, but the authors suggest that this it would not be used since librarians tend to favor the more passive "armchair" strategy, typically favored by educators. They found these results are even more disturbing because they represent the views of the more highly-paid librarians, the administrators, which implies that technological awareness is not important for professional success. Do administrators in the US fall into that same trap? (KJP)

FROM ALPHA TO OMEGA

Pasterczyk, Catherine. Mathematical notation in bibliographic databases. *Database*. 13 (4): 45-56; 1990 August.

Representing and retrieving mathematical symbols in an online database search can be a challenge, not only when searching the literature of mathematics but occasionally also in chemistry, physics and engineering. Searchers need to be familiar with symbolic notation in order to understand the search topic and improve recall. Translation of symbols in search statements is hindered by the most databases' stopwords, since the addition of "by" or "of" to a definition can add necessary precision to search results. And the same symbol will have different meanings in chemistry and mathematics. Searchers are cautioned to search all variations of a symbol, including how it is spoken, to consult the printed database instructions and standard mathematical texts, even to consider including verbal translations of symbols with totally different meanings but which look very similar, to allow for inputting errors. The seven pages of appended tables which illustrate mathematical notation as used in MathSci, Inspec, Compendex and Chemical Abstracts, and bibliographic references to sources of information on mathematical symbols make this a very useful reference for searchers in the sciences. (KJP)

IEEE, FULL TEXT, ONDISC

Stackpole, Laurie, H. CD-ROM in a federal scientific-technical library. *CD-ROM End User*. 2 (2) 60-62; 1990 June.

The Ruth H. Hooker Library offers four very different CD-ROMs to the 1500+ researchers at the Naval Research Laboratory. Applied Science and Technology Index is not used heavily, probably because the information they need is more likely to be found in online databases such as Inspec and Chemical Abstracts, and they are satisfied with searches on these databases done by members of the reference staff. They also have PC-SIG, which offers free shareware that users can download for use on their own PC's. The other two CD-ROMs are full text databases. Computer Library from Ziff Communications Co. offers full text access to 10 computer publications, as well as abstracts of articles in more than 120 related computer and electronic publications. But the one that has generated the most excitement is the IEEE/IEE Publication Ondisc (IPO) which contains citations plus the entire text, including all graphics, to journals and conference proceedings since 1988 from the Institute of Electrical and Electronics Engineers and the Institution of Electrical Engineers. Users prefer this CD-ROM to a Dialog search because they can play around with it and "perform chancy searches." This full text, low stress approach to information retrieval may fulfill the function that designers of online databases have been promising for decades. IPO is still in beta test, but many are anxiously awaiting its arrival on the market. (KJP)

T - #0250 - 101024 - C0 - 212/152/9 [11] - CB - 9781560241355 - Gloss Lamination